D0536656

Petroleum and
Global Tectonics

Petroleum and Global Tectonics

Edited by Alfred G. Fischer

and Sheldon Judson

Princeton University Press

Library of Congress Cataloging in
Publication data will be
found on the last printed
page of this book

This book has been composed in
Linotype Times Roman

Printed in the United States of America
by Princeton University Press,
Princeton, New Jersey

Contents

Prologue

The papers of this volume were presented at the 109th meeting of the Princeton University Conference held in Princeton on March 10 and 11, 1972. Sponsored by the Department of Geological and Geophysical Sciences at Princeton University, in honor of Hollis D. Hedberg, the conference was attended by 140 representatives from government, industry, and the academic world.

We initiated this conference because of our belief that the symbiotic relation between the academic world and the world of affairs, vital to both, ever requires cultivation. Periodically, developments in one field or the other demand common discussion. In this case the academic concepts of plate tectonics seem to offer solutions to many long-vexing puzzles in geology, and force a re-examination of a wide spectrum of geological thought. The processes called on in plate tectonics must bear directly or indirectly on the formation and accumulation of petroleum. It was our hope to bring together practicing geologists of the petroleum industry with their more cloistered colleagues of the educational area: to examine together the significance of current tectonic thinking to the search for petroleum, and conversely to consider the significance of the data gathered by industry to the less pragmatic activities of the academic earth scientist.

The specific subject seemed appropriate not only because it appeared timely but also because of the role of Princeton geology on both sides of the professional fence. The landmark syntheses by Hess in 1960 and 1962 set the stage for recognition and acceptance of the process of sea floor spreading, a term coined by Robert Dietz in 1961. This was followed by the work of Fred Vine on magnetic anomalies of the sea floor and their use as indices to the rate of sea floor movements—work that was begun at Cambridge and carried on during his years at Princeton. J. Tuzo Wilson, a Princeton Ph.D. of 1935, added to our understanding of the mechanics of the process with his 1965 paper on transform faults and has continued to be an eloquent spokesman for the theory. In 1968 Jason Morgan of Princeton (almost simultaneously with Xavier Le Pichon) defined the world's major plates and their motions. He followed this in 1971 with the suggestion that a few convection plumes are the driving force in the movement of plates. Thus, much of the new tectonics has had Princeton roots.

Beyond this, Princeton has sent many of its graduates into the petroleum industry: of an estimated 300 Princeton-trained geologists who have built their careers in this segment of our profession, some 200 hold Princeton B.S. or B.Sc. degrees and 100 hold Princeton Ph.D.'s.

From these chauvinistic comments arises another observation, both more general and more important. Where else but in an unhurried academic world can persons be dedicated, for years, to such questions as: Why the young age of the sea floor? Why the alternating stripes of magnetic intensity on the ocean's floor? Or what is the significance of the deep ocean trenches? But if the questions are asked within ivied walls, the techniques and the data for their solution are assembled in greatest abundance in the worldwide network maintained by industry in its search for fuel and mineral resources, and in the government supported surveys on land and sea: each of these areas of activity depends on the others.

Of course, effectiveness and progress come down eventually to the individual. In each arena there are a few who stand out. One of these has been Hollis Hedberg, in whose honor this symposium was organized and these proceedings assembled. A deep interest in the earth and all of its manifestations, a penetrating knowledge of the history of earth science, an insistence on logical clarity in preference to fashionable or traditional modes of thinking, an ability to ask incisive questions, the patience and doggedness to gather the pertinent facts, and a forthright personality with a friendliness in depth: these attributes have forged the remarkable career to which Princeton's former President, Robert F. Goheen, refers in subsequent pages. Vigorously active in both the academic and industrial fields, Hollis Hedberg has brought increased vitality to both.

We delighted in the cooperation of our authors, not only in their wholehearted participation as principals in the conference but also in providing us with the manuscripts for the papers which follow.

In addition, several people participated as moderators and panelists during the conference. Their remarks provided the glue and the focus for the general proceedings and we thank them here. Creighton A. Burk, Mobil Oil Corporation, served as chairman of the proceedings on the first day and John F. Mason of Continental on the second day. Panelists on both days of the conference were Philip H. Abelson, Carnegie Institution of Washington; Albert W. Bally, Shell Oil Company; Howard R. Gould, Esso Production Research Company; Melvin J. Hill, Gulf Oil Corporation; John M. Hunt, Woods Hole Oceanographic Institution; and Lewis G. Weeks of Weeks Natural Resources.

We acknowledge the Princeton University Conference faculty committee, chaired by Edward A. Frieman, for giving the conference its approval. The Conference Office provided excellent logistics and we mention with pleasure Edward P. Hubbs, the Administrator of that office; William H. O'Brien, Jr., Assistant Director; and particularly Marion Stark, Assistant to the Director.

Finally, to Princeton University Press, represented by its Science Editor, Mr. John Hannon, and to Gail Filion we are indebted for the excellent editorial assistance in the production of this volume.

THE EDITORS

Welcome to Conference Participants

Ladies and Gentlemen:

I know that I speak for my Princeton colleagues as well as myself when I say that you who have come to join with us in honoring Professor Hollis Hedberg by participation in this one hundred ninth meeting of the Princeton University Conference have our most cordial welcome. Beyond that, we hope that your stay with us will be enjoyable, satisfying, and productive of exchanges of ideas of value to all involved.

"Petroleum and Global Tectonics" —that is quite a conference title for a sometime classicist—more earth-bound than earth-wise—to face at 8:45 in the morning.

You will forgive me if I attempt no learned exegesis on the beauty and significance of Wilson's theory of transform faults, Morgan's plume theory, or anything like that. Yet, that word *tectonics* is one which I can grasp. I find in it a peculiar, most admirable, twofold appropriateness for this conference.

The dictionary tells me that *tectonics* is the branch of geology that deals with the earth's structure, especially folding and faulting, and I take it that this is the substantive area of your discussions. But the dictionary also reminds us that the term tectonics is derived from the Greek word *tektōn,* meaning "joiner," "craftsman," "builder." And it is on this more ancient meaning that I would pause for a moment.

We are here both to advance knowledge and ideas and to honor Dr. Hollis Hedberg. He truly is *tektonikos*—indeed *tektōn sophos*—that is to say, a master joiner and builder. In the course of a lifetime of distinguished service to his discipline, service which has earned him international recognition and honor, he has attained first-rank standing in academic circles while simultaneously pursuing a successful career in the petroleum industry, most notably, since 1946, with Gulf Oil as a senior officer concerned with exploration. In short, at once a man of thought and a man of affairs, he has built bridges between the theoretical and the practical, contriving lines of communication that reach from the realms of teaching and research to provide a basis for decision making and action in the world of affairs. So it is very fit that Professor Hedberg should be honored by a conference on tectonics, and it is a great pleasure for me on behalf of the University to join in saluting him.

We here at Princeton are proud of the contributions that this University has helped make to the understanding of global tectonics, in their more technical sense, through the work of men such as Hess, Vine, Wilson, and Morgan. As you know better than I, their investigations and theories have had a potent impact on modern geology and, in the last dozen years, have substantially changed our ideas about the physical processes and history of the earth.

Global tectonic theory, a product of academic inquiry, has the potential, I am told, powerfully to influence practical affairs. Moreover, Professor Judson has indicated to me that, through this conference, he and his academic colleagues hope to enlist scientists and engineers of the petroleum industry in a shared effort to find the ways in which the new concepts of global tectonics and the resulting increased understanding of the mechanisms of accumulation of earth resources can best be utilized for the benefit of mankind.

This is the sort of master joiner's work that Hollis Hedberg understands. To the extent that you are successful as bridge builders of this sort, joining together the academic and the practical, to that extent will you honor Professor Hedberg in a meaningful and long-lasting way. I wish you well in these efforts, and I thank you for coming to Princeton.

ROBERT F. GOHEEN

PETROLEUM AND
GLOBAL TECTONICS

Our series of papers is appropriately begun by one of the Elder Statesmen of plate tectonics—Sir Edward Bullard, whose Department of Geophysics at Cambridge, England, has been one of the focal points at which the geological revolution commenced: Specifically, where, in 1965, he, Hess, Wilson, Vine, Matthews, and McKenzie had ample occasion to exchange thoughts after the Midoceanic Ridge system had been recognized as a site of lithosphere generation. Sir Edward brings to this group not only a distinguished background, an intimacy with the development of plate tectonics and its protagonists, and a facile pen, but also the perspective of a classical physicist.

Overview of Plate Tectonics

Edward C. Bullard[1]

INTRODUCTION

The new ideas that have come into geology during the last fifteen years are not merely an addition to the previous stock of knowledge about the earth, they represent a profound change in attitude. The tradition of geology was to grow by the addition of detail. The accumulation of this detail resulted in local history, that is, in an account of the sequence of sedimentation, erosion, folding, metamorphism, and igneous activity to which an area has been subjected. The detail did not, however, produce a global theory comparable to those of physics, chemistry, or molecular biology; in fact the detail became a burden to be borne rather than a window through which to see a general plan. This is perhaps an extreme view and an exaggeration; there was a time when the development of geological theory was at the centre of scientific and public interest. The development of the ideas of the evolution of living things in the nineteenth century was, in large part, a geological enterprise; evolution was a truly general theory comparable

[1] Department of Geophysics, University of Cambridge.

in width and in predictive power to the theories of physics. Similarly the controversies of the Neptunists and the Plutonists were genuine theoretical discussions comparable to those between the supporters of Ptolomy and Copernicus (except that in the geological dispute both sides were in large part wrong).

In the first half of the twentieth century, geology was no longer at the centre of scientific interest; the discovery of the subatomic world and the great intellectual achievement of quantum mechanics drew off the brightest young men and most of the money. Geologists became a somewhat isolated group cut off by their mounds of detail and their repulsive nomenclature from the mainstream of science. The geophysicists and the geochemists nibbled around the edges of the subject, but the central citadel of tectonics and stratigraphy remained largely untouched.

The new ideas have come primarily from a study of the geology of the ocean floor. It is remarkable that for so long there was so little interest in the two thirds of the earth's surface that is covered by water. To confine consideration to a third of the object being investigated is clearly imprudent, unless one is very sure that the one third that is studied is a fair sample of the whole, which it proved not to be. Why was this clearly irrational strategy followed so universally and for so long a time in the face of the strongest hints that the oceans tell a different story from the continents? There is no simple explanation; a complex of sociological and technical factors combined to make it so. First, and perhaps most important, most geologists believed all general theories to be ill-founded, almost immoral. If one does not believe it is possible to discover the general processes of which the observed geological features are the expression, then one can only study detail for its own sake and it is easier to do this in New Jersey than in the central Pacific. Furthermore, the outdoor, natural history aspect of geology had attracted mostly quiet, modest men whose image of themselves was not that of the tycoon manqué battling in Washington for money, ships, and explosions. Unlike the physicists, they did not have the experience of using the unlimited funds unlocked by wartime successes.

The critical time came in the years after the war. The foresight of a few men, particularly Vening Meinesz and Richard Field, had directed the attention of a few young physicists and geologists, among them Harry Hess, Maurice Ewing, and Roger Revelle, to the possibilities. Wartime experience had given confidence and know-how, money was available, and, almost without realising it, we had entered a new phase in the study of the earth.

It is to be expected that a great increase in knowledge and the development of new theories in a subject will react on its practical applications. The purpose of this meeting is to discuss the relevance of the new views of geological processes to the search for oil. This introductory paper is intended to provide a background and to say, in outline, what the new ideas are. There have been so many reviews in the last few years that a full account with a marshalling of all the evidence seems hardly necessary. References to a few recent discussions are given at the end of this article.

THE SEA FLOOR

The primary source of the new ideas has been the study of the floor of the deep ocean beyond the continental edge. The first discovery was that the ocean floor is a new world that can not be discussed in terms of the relatively well-known geology of the continents. Everything is different. On land most mountains are carved by water and ice from folded and metamorphosed sediments and granitic igneous rocks. At sea almost all topography is volcanic and almost all igneous rocks are basalts.

Most significant of all is the difference in ages. On land, rocks are of all ages from 4000 million years onwards. Only the record of the first 10% of the earth's history from 4500 to 4000 million years is totally missing. The very oldest rocks are rarely found, but great areas are occupied by rocks of ages exceeding 2000 million years. In the deep sea beyond the continental shelf hardly anything older than the Cretaceous—that is, older than 150 million years—has been found, either on the surface or in the cores obtained by drilling. This was a truly remarkable and unexpected result, about which most of us were at first doubtful. One would expect the sea floor to be covered with a veneer of recent sediments even if older rocks lie below. Conviction came with the realisation that lower Tertiary and Cretaceous rocks are often found, but never anything older than Jurassic. The obvious importance of the matter led to a considerable effort in dredging on fault scarps and more recently to the JOIDES drilling project. It is now certain that the floors of all the present ocean basins have been formed in the last 4% of geological time. No doubt this will be found not to be literally true, somewhere we shall find earlier rocks preserved by some chance in some corner of the ocean. The main result, however, is beyond doubt: the continents are many times older than the ocean floor. This conclusion poses a double problem: if the ocean floor is young, it must have been formed recently. Furthermore, if there have always been oceans, there must exist a mechanism for disposing of their floors. We look, therefore, for processes of creation and destruction of ocean floor.

THE CREATION OF OCEAN FLOOR

The most striking feature of the ocean floor, perhaps the most striking topographic feature of the earth, is the great chain of submarine mountains, the midocean ridge. The continuity of the ridge is now well established; it runs from near the mouth of the Lena River in Siberia, across the Arctic Ocean, through Iceland, down the whole length of the Atlantic and around South Africa into the Indian Ocean. Here it bifurcates— one branch runs north to the Gulf of Aden and the Red Sea, and the other runs southwest, between Australia and Antarctica and into the Pacific. Here it turns north and runs into the Gulf of California; a final section starts in northern California at Cape Mendocino and runs north parallel to the coast of Canada.

For most of its length, indeed everywhere except in the Pacific, the crest of the ridge is cut by a steep-sided valley typically 30 km or so in width. This valley is seismically active, a line of earthquake epicentres being accurately aligned along it. These earthquakes show that in each shock the valley is cracking open. Rocks dredged from the floor of the valley are recent lavas, which suggests the not surprising conclusion that the cracks are filled by basalts rising from below. As would be expected, the heat flow from beneath the valley, through the ocean floor, is usually several times that found away from the ridge axis.

Here in the central valley we see a process of ocean floor formation. The valley floor cracks, new lava comes up, and a new strip of ocean floor is created. This is the process of "ocean floor spreading" first envisaged by the late Harry Hess in 1962 though perhaps dimly foreshadowed by Arthur Holmes in 1929. The detailed progress of the spreading is recorded on a double tape-recorder. The lava extruded on the valley floor is magnetised in the direction of the earth's field at the time of cooling. Thus a strip of floor all magnetised in the same direction is gradually built up. When the field reverses, as it does at random intervals

of between 100,000 and a million years, a strip of the opposite polarity
is built up. Thus, as the spreading progresses, ocean floor is formed
on both sides of the central valley in strips of alternating polarity. The
strips can be regarded as isochrons. The whole length of a given strip
was formed on the crest of the ridge at a time which is known if the
timetable of the reversals is known.

This idea, which was originated by Vine and Matthews in 1963, is
a global theory of a kind very uncommon in geology. It is interesting
to note, as an example of the prejudice against such things, that L. W.
Morley who had the same idea at about the same time was unable to get
his paper accepted for publication. The theory is important because it
makes predictions that can be verified. It predicts the form, but not the
scale, of the magnetic profiles that will be obtained across the ridge anywhere in the world, using only the timetable of reversals established by
the quite independent evidence of palaeomagnetism and radioactive dating. It also predicts what will be found in drilling on a particular magnetic
stripe. Just beneath the ocean floor there will usually be recent sediments, though they may have been removed by erosion by currents;
beneath this will be sediments of increasing age back to the time when
this piece of sea floor was on the axis of the ridge, back, that is, to
the age of the stripe determined from the chronology of the magnetic
reversals. Beneath the sediment will be lavas of the same age as the
basal sediments. These predictions have been beautifully verified in
numerous holes drilled by JOIDES in the Atlantic and Pacific Oceans.

If two plates are moving apart on the surface of a sphere then a
theorem proved by Euler in the middle of the eighteenth century shows
that their relative motion may be regarded as a rotation about a vertical
axis through a point called the pole of spreading. This theorem elegantly
describes the geometry of the motion and leads to verifiable consequences; for example, the rate of separation of the two plates must be
small near the pole of spreading and a maximum 90° away. Another
prediction concerns the offsets of the ridge axis. The crest of the ridge
and its central valley do not form a continuous, unbroken line; they
are displaced by cross faults along which a transverse sliding motion
is taking place. If these faults are not to open they must lie along circles
of latitude relative to the pole of spreading. All such predictions are
very well verified by a study of the topography, the earthquakes, and
the magnetic lineations.

The result of all this work is a rather detailed picture of the motion of the plates today and of its changes in the past back to the beginning of the Cretaceous.

THE DESTRUCTION OF OCEAN FLOOR

The rate of creation of new ocean floor on a ridge axis is typically 2 to 20 cm/yr. Such a rate must be regarded as rapid on a geological time scale. A velocity of 20 cm/yr would result in a displacement of 20,000 km in a 100 million years; it is clear that there must be some process for disposing of the ocean floor and returning it to the upper mantle, from which it emerged as igneous intrusions and lava on the ridge axis.

The obvious sites for such a destructive process are the belts of earthquakes that follow many of the continental edges and which are often associated with island arcs, ocean trenches, and andesitic volcanoes. The most conspicuous of these belts is that around the Pacific but there are others in Indonesia, in the Caribbean, and in the Scotia Arc between South America and Antarctica. The earthquakes lie on inclined surfaces with shallow foci on the seaward side, near the ocean trenches, and deep ones under or on the landward side of the island arcs. The deepest earthquakes lie at a depth of between 600 and 700 km.

It seems that near the ocean trench the plate dives down at an angle of about 45° to the horizontal and forces its way into the upper mantle. The cold plate will slowly be heated by conduction from the surrounding hotter rocks of the mantle. The time taken for it to soften will depend on its thickness, which is only vaguely known. Something around 50 to 100 km seems likely; this would be compatible with a loss of strength by the time it reached the depth at which earthquakes cease. The picture of a descending plate has been confirmed by showing that earthquake waves travel more quickly down the cold plate than they do through the surrounding hotter rock.

PLATE TECTONICS AND CONTINENTS

Plates have three kinds of boundaries: midocean ridges where they are born, transform faults where they slip past each other, and trenches where they die. A map showing the distribution of earthquakes is the key to the system, such a map delineates the plates as they exist today and suggests a wide range of questions about their motions and past history. It shows, as was long ago pointed out by Eduard Suess, that there are two kinds of coast line. At one kind a plate is being destroyed, as off the east coast of Japan; at the other kind there are no earthquakes, and it seems that no motion is taking place between the ocean basin and the neighbouring continent. The eastern coast of North America is a typical example of such a quiescent coast, where there is no plate boundary and the continent and the ocean floor move together as a single plate. This is the old idea of continental drift but with the emphasis shifted; the continents no longer plough through the ocean floor, they are carried along with it. This change comes directly from the study of marine geology; the mechanism of plate creation is in the ocean and that of plate destruction at the boundary between ocean and continent. Neither could be understood till attention was given to the ocean floor as well as to the continents.The idea has, however, many consequences for continental geology.

If the Americas are moving away from Europe and Africa, then in the past they must have been closer together than they are now. At 2.5 cm/yr, a distance of 5000 km would be covered in 200 million years, it is therefore not unreasonable to suppose that the Atlantic was closed in the early Mesozoic. This apparently rash extrapolation is supported by the details of the magnetic lineations and by the JOIDES drilling. It is therefore necessary to take seriously the possibility that North and South America, Europe, and Africa were once joined in a single land mass. If they were so joined, and if the edges have not since been greatly eroded or built out, then is should be possible to fit the continents together like the pieces of a jigsaw puzzle. The edges of the pieces should be taken not at the coastline but at the outer edge of the continental shelf, which is the true edge of the continent. This idea is, of course, not new; the fit was part of the evidence used by Wegener and du Toit in their arguments for the reality of continental drift. It is curious, however, that it was not taken more seriously. Wegener's book gives only rough sketches, du Toit's map of the fit of South America and Africa is better, but is still only a small diagram. The reason for this casual approach is far from clear; Wegener and du Toit were involved in a difficult and acrimonious controversy on whose outcome their scientific reputations largely depended. Why were they so slap-dash? We shall probably never know; perhaps they thought the continental edge was certain to have been distorted by erosion, perhaps the climate

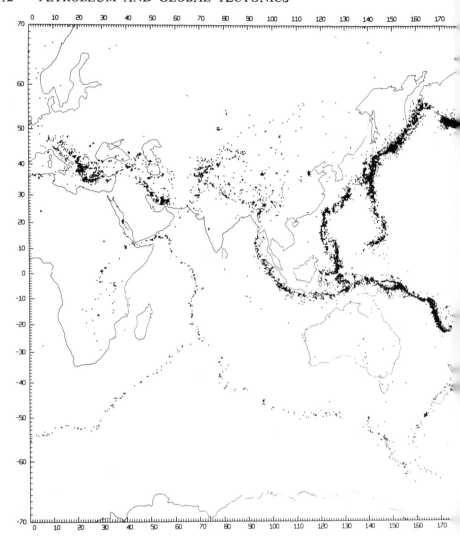

Figure 1. Seismicity of the Earth, 1961–1967, ESSA, CGS Epicenters.
(Depths 0-100 km.)

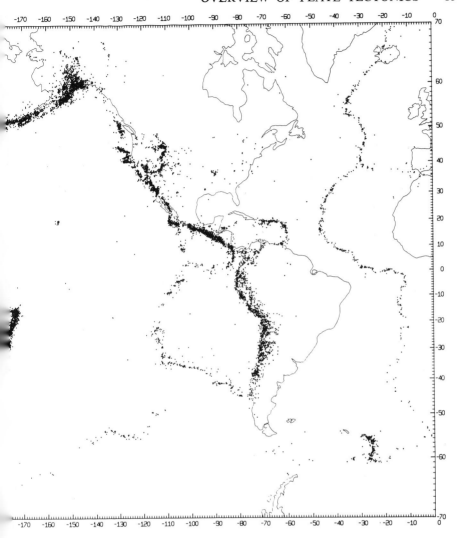

of opinion and the belief that no simple idea could work in geology
was too strong.

In fact, the fitting is not a completely trivial problem, the topography
is specified by maps on various projections, but the fitting must be done
on a sphere; it is not legitimate simply to cut out South America from
a Mercator chart and place it against Africa. The effort involved in making a careful fit is amply repaid, since it suggests many doubtful points
for further examination. The fit of South America and Africa is good
(standard deviation 90 km); the only conspicuous misfit is the Niger
delta, which has been formed since the split and would therefore be expected to overlap South America when the pieces are assembled. The
fit of North America, Greenland, and Europe is even better; that of
North America and Africa is less good and it is an important question
of palaeogeography to determine if they were ever really in contact.

The Atlantic fit gives a new importance to many geological details.
If the fit represents a real reconstruction of past geography, then not
only must the pieces fit geometrically but the geologic maps on the two
sides must also fit. The sharpness of this requirement is blurred by the
two coast lines being separated by the continental shelves whose geology
is often unknown. In some places a critical comparison is possible as
in West Africa, where the junction of the 2000 million year old rocks
with the adjacent 500 million year old rocks to the east is precisely
repeated in the expected place in Brazil. Here both continental shelves
are narrow, and the geological fit is a welcome confirmation of the
geometry.

In making the Atlantic fit, some rather arbitrary adjustments have
been made; being arbitrary they reduce the force of the argument for
the former unity of the continents around the Atlantic. From another
point of view, however, they can be regarded as predictions to be verified
or disproved by other evidence; if they are verified they are no longer
weaknesses but predictions which greatly strengthen the case. The principal examples of such arbitrary adjustments in the fit are the abolition
of Iceland, with the implication that it has no continental basement,
the retention of the Rockall Bank as a continental fragment, and the
rotation of Spain to avoid an overlap with Africa and to close the Bay
of Biscay. For all these there is now good independent evidence. The
details of the opening of the Atlantic can be followed by a study of
the magnetic lineations; the whole story is not yet completely known,
large uncertainties remain in the Caribbean and in the way in which
the Arctic Islands join northwest Greenland.

Similar attempts have been made to assemble the continents around the Indian Ocean into the hypothetical Gondwanaland. Here the fit of Australia to Antarctica is outstandingly good, but there are ambiguities elsewhere, particularly in the former position of Madagascar.

A continent carried along with a plate will ultimately arrive at a trench where the plate is being destroyed. The continent is made predominately of granite and will be unable to sink into the much denser upper mantle. If the trench borders a continent, then the approaching continent will collide with the one behind the trench. When this happens, great distortion of both continents and of the intervening sea floor is to be expected. A rather clear case of this kind is the collision of India with the southern shore of Asia. The formation of the Himalayas may well be a consequence of this collision. The details are far from clear, in particular, it is not known where the suture line lies, it may be under the Gangetic Plain, or in the mountains, or perhaps part of India has slithered under Asia and raised the Thibetan plateau. The wide area over which the earthquakes of Thibet and Sinkiang are spread is consistent with this last possibility. Another collision area is the Mediterranean, where Africa is moving northward against southern Europe.

A rather different process of mountain building is in progress in other places where there is no collision of continental masses. For example in South America the plate spreading from the East Pacific Rise is plunging beneath the Andes. It seems likely that, as the plate goes down, material is scraped off it and piled up in a chaotic mass against the edge of the continent; the great chain of andesitic volcanos that lies further inland is probably formed from material that has migrated upwards from the slowly melting plate.

The process of continental splitting can be seen in action today in the Red Sea, the Gulf of Aden, and the Gulf of California. It seems that a valley is first formed, perhaps by faulting or stretching of the crust, and that this is intermittently connected to the sea. Great thicknesses of evaporites may be formed in the valley; then, as a more permanent connection with the ocean is established, marine sedimentation starts. The valley or narrow sea widens by splitting along its axis, and gradually an ocean is formed.

A central problem of earth-history is to decide whether the break up of the continents in the early Mesozoic was a once-for-all fragmentation of a primeval continental mass, the *Pangea* of Wegener, or whether continents have been splitting, drifting, and colliding all through geologic time. Since it seems that the early ocean floor has been totally destroyed, this is not an easy matter to decide. Perhaps the best evidence comes from the study of Palaeozoic mountain ranges. The Appalachian-Caledonian system does not differ essentially from the Tertiary mountain systems and it is natural to suppose that it was made in the same way; that is to say that all through the Palaeozoic there was an Ur-Atlantic which closed at the time when the mountains were formed or shortly after and then opened again along a somewhat different line to form the modern Atlantic, which cuts the old mountains into American and European sections. Again it seems unlikely that the Ural Mountains were formed in the middle of a continent and much more probable that Asia was split at the time the sediments of the mountains were deposited. There is some palaeomagnetic evidence that Asia is a mosaic formed by the putting together of formerly separate fragments. Although very little is known of pre-Cretaceous plate movements, it seems likely that the process has been going on through all or most of the history of the earth and is not a recent innovation.

THE MECHANISM OF PLATE MOVEMENT

In talking about plate movements, we are close to the solid ground of facts observed at the surface of the earth; in talking about the forces driving the motions, we are speculating about the largely unknown interior. Statements about mechanism are therefore to be regarded as possibilities for detailed formulation and computation rather than established theories. The ultimate source of the energy of the motions is likely to be radioactive heating; the forces would then be gravitational and associated with differences of density due to differences in temperature.

Within this general framework there are several possibilities. The simplest is the so-called Rayleigh-Benard convection, which is an instability produced by heating of the mantle; to find the condition for instability is relatively easy but to find the form of the motion is not. The change of viscosity with temperature, which is very imperfectly known, plays an essential role; if the material of the mantle is plastic or behaves as a non-Newtonian fluid, further uncertainties are introduced. In fact the whole theory of Rayleigh-Benard convection may be irrelevant and the form of the motion may be determined by conditions near the surface of the earth; in particular, differences in the vertical distribution of radioactivity between the oceans and continents may play an essential part. It has also been suggested that, as the downgoing plate is colder and denser than the surrounding mantle, it will sink and pull the rest of the plate after it like a table cloth slipping off a table.

It is likely that none of the simple views is sufficient and that a satisfactory theory will be complicated. Fortunately, it is not essential, for the purposes of this conference, to understand the mechanism behind the movements.

PLATE TECTONICS AND PETROLEUM EXPLORATION

Petroleum exploration is largely concerned with the stratigraphy and structure of sedimentary basins. Its main interest is, therefore, in the vertical movements which allow, or perhaps are caused by, the accumulation of great thicknesses of sediments. Plate tectonics, on the other hand is, in large part, an account of horizontal movements. The difference is the main reason for the rather weak impact of plate tectonics on petroleum exploration. It may, however, be expected that the next stage in the development of the theory will be concerned with the processes near continental margins and will include an explanation of the formation of various kinds of sedimentary basins. It is reasonable to suppose that an understanding of the processes at work in basins will be of use in their exploration.

A number of other papers in this collection treat the classification of basins in terms of plate tectonics. Here we merely mention the principal types. The greatest volume of sediments is probably that accumulated on sinking ocean floors behind island arcs. Of such basins, the Sea of Japan is typical. The reason for the sinking is not clear, it is perhaps in some way connected with a downgoing convection current produced in the upper mantle by the cold sinking plate. The ultimate fate of such basins is a matter of speculation. If the character of the plate margin changes and the coast ceases to be bordered by a trench and a downgoing plate, sedimentation will continue and the sea will gradually be filled. The Black Sea may be an example. Continued sedimentation will blanket the oceanic basement and raise temperatures; this may produce metamorphism of both basement and sediments. The final result might be new continental crust.

A completely different kind of basin is formed in the early stages of continental splitting, as in the Red Sea. Here there are great thicknesses of evaporites covered by marine sediments. A number of other types are possible, for example the masses of sediments piled up on the landward side of a downgoing plate and the sediments formed between colliding continents as in the Ganges basin or the Mediterranean.

There is a large element of speculation in such classifications as can be seen by considering specific examples. Was the North Sea an "inland sea" behind a Palaeozoic island arc or is it a result of an abortive split in which England started to accompany America in its departure from Europe? The Gulf of Mexico is an inland sea behind an island arc (the occurrence of evaporites in deep water suggests that it was once isolated from the ocean), but what is its relation to the great thickness of Tertiary sediments on the continental shelf and on land? It is well worth going systematically through the evidence concerning such basins and attempting to fit them into a larger pattern. To do so requires a detailed knowledge of the stratigraphy and must be left to the experts on the areas.

Amid the many doubts and in spite of the absence of important results concerning the detailed relation of oil-bearing basins to plate tectonics, it is perhaps worth pointing out that there clearly is a relation. The greater part of the world's oil comes from the neighbourhood of the boundary between the African and Eurasian plates. This boundary runs from the Azores to the north coast of Africa, then along the northern margin of Africa, crosses the Mediterranean in a way that is not entirely clear and runs through Turkey and across the Middle East into Pakistan. For the greater part of its length it represents the line along which Africa and Arabia are pushing north and north east against the Mediterranean and the mountains of Turkey and Iran. It would be of the greatest interest to understand the relations of the plate edge and the motions to the basins and the oil accumulations. In particular, why are there salt deposits of such a great range of ages in Iran? Are they formed in the first stages of an opening ocean or in the last stages of a closing one? Such questions are easy to ask in general terms but very hard to answer in detail. It is a large part of the purpose of this conference to consider them.

REFERENCES

It is impossible to give here a guide to the enormous literature of the past few years on plate tectonics. There are a number of semi-popular accounts including:

The Ocean, 1969, a Scientific American Book. (also *Scient. Amer.* Sept. 1969)

Horsefield, B. and Stone, P. B., 1971, *The Great Ocean Business.* London: Hodder & Stoughton.

More technical accounts with extensive bibliographies will be found in:

Bullard, E., 1968. Reversals of the earth's magnetic field, *Phil. Trans. R. Soc.* A, 263, 481–524.

Robertson, E. C., 1972. *The Nature of the Solid Earth,* sections 13–17. New York: McGraw-Hill.

McKenzie, D. P. and Sclater, J. G., 1971. The evolution of the Indian Ocean since the late Cretaceous, *Geophys. J. R. astr. Soc.,* 25, 437–528.

The last of these papers is a striking example of the use of magnetic lineations, earthquake epicentres and fits to elucidate the history of an ocean where there has been a large change in plate motion during the Tertiary.

Heat generated or stored in the earth's interior is dissipated through the crust and lost to space, in the process termed *heat flow*. The distribution of heat flow patterns over the earth's surface contains important clues to processes occurring at depth. But heat flow is also significant in terms of the geological work it accomplishes: As sediments accumulate in basins, for example, the accumulation is progressively heated from the bottom. The role of geothermal gradients in diagenesis, compaction, and deformation of sediments is surely a large one, though poorly understood, and their importance in petroleum generation and accumulation is discussed in subsequent papers by Erdman and by Klemme. Jason Morgan, known chiefly for his role in the recognition and delimitation of the world's major tectonic plates and for his concern with plume convection in the mantle, here turns his attention to heat flow in aging oceanic lithosphere. A comparison of theoretically deduced heat flow regimes with such regimes adduced from conventional measurements shows major discrepancies: Much of the heat loss from juvenile lithosphere must occur by convective transfer (hot springs), not measured by conventional means.

Heat Flow and Vertical Movements of the Crust

W. Jason Morgan[1]

The gradual sinking of the ocean floor with increasing age is the most thoroughly understood of the phenomena producing vertical displacements in the earth's crust. And yet there exists a major discrepancy in the theories relating elevation to oceanic heat flow; the observed heat flow is not large enough to produce the necessary contraction and sinking of the ocean floor. A new model is presented here which even further accentuates this difference, and it is concluded that about half of the heat loss of the ocean floor must occur by localized hot springs, and half by conduction through the surface.

We wish to use the metric units which are appearing in more and more journals; the relation between these units and those more commonly used are as follows:

$$1 \text{ HFU} = 1 \text{ microcal/cm}^2 \text{ sec} = 41.8 \text{ milliwatts/m}^2 \ (1.4 \text{ HFU} \doteq 60 \text{ mw/m}^2)$$

$$1 \text{ cal/}^\circ\text{C gm} = 4180 \text{ Joule/}^\circ\text{C kg} \ (.24 \text{ cal/}^\circ\text{C gm} \doteq 1000 \text{ J/}^\circ\text{C kg})$$

$$1 \text{ cal/cm sec}^\circ\text{C} = 418 \text{ watt/m}^\circ\text{C} \ (.007 \text{ cal/cm sec}^\circ\text{C} \doteq 3 \text{ w/m}^\circ\text{C})$$

[1] Department of Geological and Geophysical Sciences, Princeton University.

Figure 1 shows the general model that accounts for the heat loss of newly created oceanic plate. At the rise crest, the material is nearly molten right up to the top surface. The thermal gradient is very steep, and there is great heat flow. Far from the rise crest, the thermal gradient becomes very uniform—essentially a straight line connecting the 0°C temperature at the top with the constant temperature

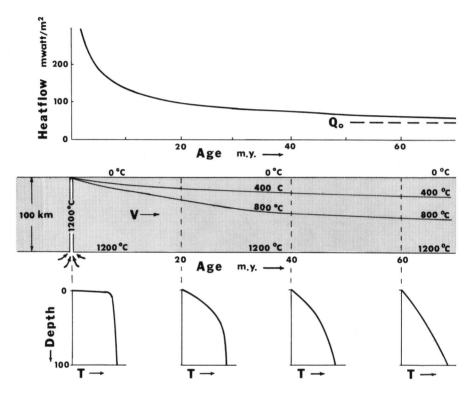

Figure 1. Schematic diagram showing how the sea floor spreading predicts a heat flow pattern because of the transient cooling of the newly formed slab.

T_0 (1200°C) specified at the bottom of the slab of thickness H (100 km). At the distances intermediate between "at" and "far from" the rise crest, the temperature profile in the lithosphere is changing with time, i.e. the pattern of high heat flow near a rise is due to the (transient) loss of heat of the cooling lithosphere slab.

The limiting thermal gradient far from the rise multiplied by the thermal conductivity, K (3.6 watt/m°C) gives the heat flow that comes from below the lithosphere, i.e. sources deeper in the mantle must supply sufficient heat to maintain the base of the slab at the specified T_0. (The "transient" heat of the cooling lithospheric slab also comes from the mantle, but it comes all at one time as upward convection at the crest as opposed to conduction upward over a broad area.) It is implicitly assumed in such a model that the material below the "slab of thickness H" is well stirred up by convection in order to maintain this boundary at a constant temperature T_0. If this lower surface cannot be maintained at a constant temperature, then the heat flow at ages beyond 30 or 50 MY will change in a manner different from the exponential decay towards equilibrium as predicted by this model.

With a limited set of assumptions, Figure 1 can be transposed into an exact mathematical model. The top and bottom surfaces are maintained at constant temperatures, 0°C and T_0, respectively. The newly created slab moves only horizontally with a uniform velocity of **v**. The effects of upward convection of material in a narrow zone near the rise crest is approximated by specifying the temperature along this boundary. Within the slab, the temperature pattern must satisfy the equation of heat transport,

$$\rho c \frac{\partial T}{\partial t} + c\rho \mathbf{v} \cdot \boldsymbol{\nabla} T = K \boldsymbol{\nabla}^2 T + A \qquad (1)$$

This equation relates the change of heat content at a small region fixed in space to the difference in heat convected into and out of this region, and the heat generated by radioactive decay (or other process, such as shear heating) within the tiny volume. We may safely omit A, the radioactive heat generation in the slab. Choices of A will affect the temperature versus depth profile in the upper 100 km, but not the time constants that will be our major concern here. In other words, the equilibrium gradient would not be a straight line as in Figure 2, but the exponential decay of the "sine" terms will be unchanged. We shall assume that the parameters ρ, c, and K (the density, specific heat at constant pressure, and conductivity) are constant and are 3360 kg/m³, 1000 J/kg°C, and 3.6 watt/m°C, respectively. If we assume a steady process, the solution to equation (1) is (McKenzie, 1967):

$$T(x,z) = \frac{T_0 z}{H} + \sum_{n=1}^{\infty} a_n \exp\left([R - \sqrt{R^2 + n^2\pi^2}] \frac{x}{H}\right) \sin\left(\frac{n\pi z}{H}\right) \qquad (2)$$

where $R = \rho c \mathbf{v} H / 2K$ is the Rayleigh number. If R is very small, then the x-dependence is governed by the $n^2\pi^2$ term which represents horizontal conduction away from the vertical boundary at the crest. For large Rayleigh numbers, horizontal convection dominates over horizontal conduction. In this case, the term expressing the x-dependence of the heatflow, $\exp([R - \sqrt{R^2 + n^2\pi^2}]x/H)$, can be accurately approximated by $\exp(-n^2\pi^2 K x / \rho c H^2 \mathbf{v})$. Calculations for a variety of Rayleigh numbers show that the "small" and "large" Rayleigh numbers occur at spreading velocities of 1 mm/yr and 3 mm/yr—i.e. for rates less than 1 mm/yr the conduction effects are large, whereas all rates greater than 3 mm/yr are practically indistinguishable from the approximation shown above. In the approximation, the horizontal dependence is expressed in terms of the ratio of the horizontal distance and the spreading rate. If we define $t = x/\mathbf{v}$ to be the "age" of a given piece of ocean floor, we see that the heat coming from the floor depends only on its "age." The other constants have the dimension of time, and we shall define the thermal time constant to be $\tau = \rho c H^2 / \pi^2 K$, and the time constant of a given mode to be $\tau_n = \tau/n^2$. Equation (2) was used in all of the calculations made with the constant coefficient model (i.e. the program included the effects of horizontal conduction), but in our discussion we shall always refer to the simpler equation for "fast" ($\mathbf{v} > 3$ mm/yr) spreading rates.

$$T(t,z) = T_0 z/H + \sum_{n=1}^{\infty} a_n e^{-t/\tau_n} \sin\left(\frac{n\pi z}{H}\right) \tag{3}$$

The variations in the temperature profile with age will express itself in two observed features, the heat flow from the top surface, Q, and the change in elevation, h.

$$Q(t) = K \left.\frac{\partial T}{\partial z}\right)_{z=0} = K \frac{T_0}{H} + \sum_{n=1}^{\infty} \frac{n\pi K}{H} a_n e^{-t/\tau_n}$$

$$h(t) = h' + \int \alpha T \, dz = h_0 + \frac{\rho_m}{\rho_m - \rho_w} \sum_{n=1,3,5}^{\infty} \frac{2\alpha H}{n\pi} a_n e^{-t/\tau_n} \tag{4}$$

The new variables introduced are α, the volumetric coefficient of thermal expansion, and ρ_m and ρ_w, the density of mantle and the density of water. In computing the changes in elevation, we have assumed that the pressure of the slab/fluid boundary at depth H is constant and that the height of the column is then inversely related to its density. In other words, the level of compensation for isostacy is at the base of the cooling slab. We have assumed that any horizontal stresses generated by thermal contraction are relieved by slow creep in the slab—hence our use of the volumetric coefficient of expansion. The ratio $\rho_m/(\rho_m - \rho_w)$ corrects for the weight of the water load changing as the depth increases.

Our problem now is to find the set of Fourier coefficients that describes the thermal profile at the rise crest—our formulas will then show how such an initial profile will evolve with time. These coefficients represent the difference between the vertical temperature profile at $x = 0$, $T(0,z)$, and the flat ramp which shows the steady state profile after all of the transient heat is gone.

$$a_n = \frac{2}{H} \int_0^H [T(0,z) - T(\infty,z)] \sin\left(\frac{n\pi z}{H}\right) dz \tag{5}$$

The simplest case is to assume that the temperature along the "crack" is identical to the constant temperature T_0 along the bottom. In this case we get $a_n = 2T_0/n\pi$. This solution is shown graphically in Figure 2 and analytically by

$$T(t,z) = T_0 z/H + \sum_{n=1}^{\infty} \frac{2T_0}{n\pi} e^{-t/\tau_n} \sin\left(\frac{n\pi z}{H}\right) + \gamma z$$

$$Q(t) = KT_0/H + \sum_{n=1}^{\infty} \frac{2KT_0}{H} e^{-t/\tau_n} + K\gamma \tag{6}$$

$$h(t) = h_0 + \frac{\rho_m}{\rho_m - \rho_w} \sum_{n=1,3,5}^{\infty} \frac{4\alpha T_0 H}{n^2\pi^2} e^{-t/\tau_n}$$

The terms γz and $K\gamma$ were added to the above solution to represent a slightly more general initial boundary condition—the case where the initial temperature along the "crack" is not constant but increases with depth according to $T(0,z) = T_0 + \gamma z$. The leftmost part of Figure 2 shows the steady state solution after all of the exponential terms have decayed. The next four parts show Fourier components of the solution, and the last part shows the successive sums of these components. Note that after only four terms, our hypothesized constant temperature T_0 along the initial crack is fairly well approximated—with more terms it would be even better represented.

There are several points we wish to note in Figure 2 and equations (6). First, only one time constant of heat decay at the rise crest is commonly referred to, whereas we see there are in fact many time constants—one for each Fourier term. In our problem where $\tau = 10$ MY, we see that after only 5 MY the dominant term shown in Figure 2 will decay by $e^{-\frac{1}{2}} = .61$, whereas the next largest term has decayed by $e^{-2} = .12$ and the other terms would be unmeasurably small. Thus, except for a region near the crest, the heat flow should fall off exponentially with age. Second, note that the height of the rise as caused by

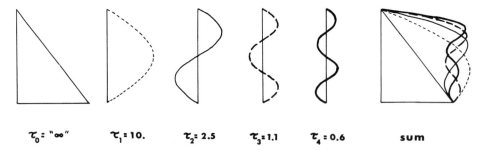

$\tau_0 = $ "∞" $\tau_1 = 10.$ $\tau_2 = 2.5$ $\tau_3 = 1.1$ $\tau_4 = 0.6$ sum

Figure 2. Diagram showing how the terms in equation (6) are combined to match the temperature versus depth at time $t = 0$. The sum of just the first 4 Fourier terms is a fair approximation to the initial temperature versus depth pattern. Notice that all of the terms have the same gradient at the top; this produces an infinite heat flow at time $t = 0$ in this simple model. Also notice that the even Fourier terms have equal amounts of "plus" and "minus"; thus only the odd terms contribute to the decay of the average temperature of the slab and thus to elevation changes.

thermal expansion should be even more perfectly represented by an exponential curve—only every other term affects the expansion (those with an odd number of half-wavelengths, the even half-wavelength terms have a self-canceling feature). Third, note that each additional Fourier term contributes the same amount $(2KT_0/H)$ to the initial heat flow at the rise crest. This is an artifact of the model; it results from our assumption that the temperature at the boundary is T_0 right up to the top surface. If we assume that the initial temperature at the crest boundary decreases to 0°C near the top, as would result for example if we ascertained that the conservation of energy was satisfied in setting the boundary temperatures (i.e., choosing the temperatures along this boundary in a way that the energy conducted and convected out of a small segment equaled the energy influx from the segment below), then the higher order Fourier coefficients would decrease and the model would not give infinite heat flow at the origin. Variations of the temperature on this boundary will be considered later in connection with Figure 6.

Figure 3 shows heat flow data from the Atlantic and Pacific Oceans. The wavy dotted line is the "50% line"; half of the observations are above this line and half below. The solid line is the theoretical fit by McKenzie (1967); the dashed theoretical line will be discussed later. The thermal decay time of McKenzie's model is very close to 10 MY, and the model parameters used to obtain this time are $H = 50$ km for the thickness of the slab and $T_0 = 550°C$ for the temperature at the base of the slab. (McKenzie assumed the spreading rates in the Atlantic and Pacific were 10 mm/yr and 40 mm/yr, respectively. The difference between these assumed values and the presently accepted values is of no importance here.) The time constant of 10 MY requires the thin "lithosphere"; the thin lithosphere then demands the low 550°C temperature at its base so that the steady state heatflow will not be excessive. A temperature at the base of 1100°C would produce the same final heatflow if the conductivity were reduced by half ($Q = KT_0/H$), but the reduction of the conductivity by half would cause the time constant to double to 20 MY ($\tau = \rho c H^2/\pi^2 K$). The density and heat capacity are the best known parameters, and so, to reduce the time constant to 10 MY, the plate would have to be even thinner ($H = 35$ km)—an endless circle. The point we wish to make is that $\tau = 10$ MY best fits the observed heat flow data at the rise crests; longer time constants do not fit the crest data nearly so well. This 10 MY time constant and the 50 or 60 km thickness for the lithosphere it implies have been widely used, but it is based only on the observed pattern of heat flow at the rise crest and ignores the problem of the very low base temperature.

The cooling lithospheric slab has another observable effect—as it cools there is thermal contraction and a gradual sinking of the ocean floor. Menard (1969), Sclater and Francheteau (1970), Sclater, Anderson, and Bell (1971), and several other authors have compared the age of ocean floor with its depth (corrected for sediment thickness and an assumed isostatic correction for this thickness). Empirical curves relating depth to age have been established. There is very little scatter in the depth/age date—some authors (Anderson and Davis, 1973) have used a depth difference of only 250 m to infer an age relation in a com-

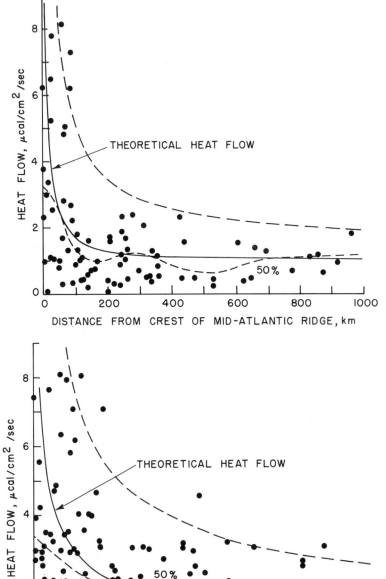

Figure 3. Heat flow measurements in the Atlantic and Pacific Oceans plotted versus distance from the rise crest. The solid line shows the heat flow predicted by the model by McKenzie (1967), the light dashed line is an empirical fit with 50% of the values above and below the curve, and the heavy dashed line is the heat flow predicted in this paper. (Figure adapted from McKenzie, 1967.)

plicated area of the East Pacific Rise.. Figure 4 is an example of an empirical fit to the age/depth data. We shall use the longer time constant, $\tau = 78$ MY, to calculate the thickness of the lithospheric slab. The shorter time constant shown here has no theoretical basis; it was empirically chosen to best fit the data. From equation (6), we see that our model predicts the next time constant in the sum, τ_3, should be $1/n^2 = \frac{1}{9}$ of the primary constant, and the amplitude of this coefficient should be $\frac{1}{9}$ of the 3904 m coefficient of the first term. The observed amplitude is very close to 1:9. The empirical time constant is less than the predicted ratio; but we could attribute this to the simplifications assumed in the linear model or we might claim that the 1:9 ratio represents the data within allowable errors. The age versus depth data have much less scatter than the heat flow measurements, and they very closely follow the exponential decay curve. We may construct a model that matches this decay with the following parameters.

$$H = 150 \text{ km} \quad T_0 = 1250°C \quad K = 3.1 \text{ w/m°C} \quad \rho = 3360 \text{ kg/m}^3$$
$$c = 1000 \text{ J/kg°C} \quad \alpha = 3.6 \times 10^{-5}/°C \quad \gamma = 0.5°C/km \tag{7}$$

This model predicts:

$$\tau = \rho c H^2/\pi^2 K = 78 \text{ MY}$$
$$Q_0 = K(T_0/H + \gamma) = 29 \text{ mw/m}^2 \tag{8}$$
$$h = \left(\frac{\rho_m}{\rho_m - \rho_w}\right) \frac{1}{2}\alpha T_0 H = 4.8 \text{ km}$$

The value of the coefficient of expansion used above, $\alpha = 3.6 \times 10^{-5}/°C$, was chosen to make the total range of sinking equal to 4.8 km. We would have preferred to use a slightly larger coefficient of expansion, $\alpha = 4.0 \times 10^{-5}/°C$. This larger value could be used and we could still obtain a total range of sinking of about 4.8 km if two additional physical processes were included in the model.

The basic assumption in relating the vertical temperature profile to the elevation was that the base of the slab and the "level of compensation" were the same. If the lower part of the cooling slab is always hot enough so that it always has the creep/flow properties that we ascribe to asthenosphere, then the level of constant "pressure" may be well above the base of the slab. That is, we are saying the asthenosphere as well as the lithosphere is involved in the cooling process—the asthenosphere at a depth of 120–150 km beneath the westernmost Pacific is

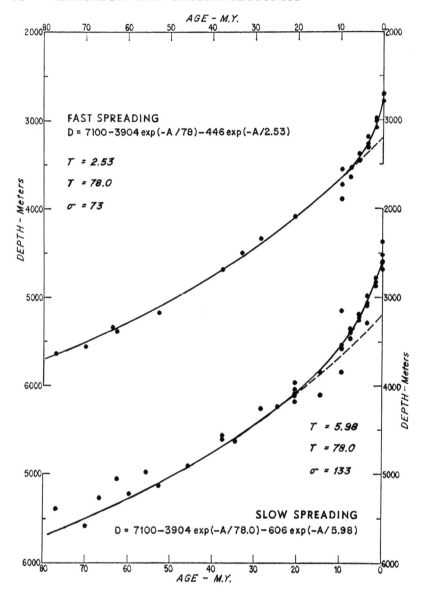

Figure 4. An empirical fit of the depth of the seafloor versus its age by Le Pichon, Francheteau, and Bonnin. The quantities T and σ are the time constants (in MY) and standard deviation (in m) of the fit. (Figure from Le Pichon, Francheteau, and Bonnin, 1973.)

slightly cooler than asthenosphere at the same depth near the East Pacific Rise. The height at which an object will float depends on the density of the fluid as well as the density of the object, and since the "old" oceans are floating on top of denser than average asthenosphere, they float higher than we might otherwise expect. Another assumption of isostacy is that the fluid below transmits no shear stresses, i.e. it is in hydrostatic equilibrium. As we assume there are important stress gradients in the asthenosphere involved in moving the plates, an investigation of this problem would require equations coupling the thermal and mechanical properties—a study which we are not prepared to make. As an example, Schubert and Turcotte (1972) show that if the return flow from descending slabs back to spreading rises is all channeled into an asthenosphere only a few hundred kilometers thick, then there must be a strong pressure gradient in the asthenosphere to drive the return flow—with the consequence that the ocean floor near the trenches would be elevated with respect to the ocean floor near the rises.

There is another factor which can reduce the 5.6 km range of sinking: the amount of radioactivity in the slab. If we assume that the upper 10 km of "crust" contains a radiogenic heat production of $A = .500$ $\mu w/m^3$ and that the lower 140 km of the slab has $A = .025$ $\mu w/m^3$, then this produces an additional heat flux of 8.5 mw/m^2 for a total heat flux out of the very old slab of $29 + 8.5 = 38$ mw/m^2 ($= .9$ HFU). This also raises the average temperature of the equilibrium gradient by 22°C, which lessens the range of temperature change in the cooling slab, but only by 3%. In order to let predicted topography match the observed range of topography, we have chosen $\alpha = 3.6 \times 10^{-5}/°C$. This choice is in effect a "fudge" that combines the effects of adopting a different thermal expansion coefficient, the effects of radioactivity in the slab, and, what we presume to be the main cause, the effect of the asthenosphere cooling as well as the lithosphere.

Figure 5 shows the heat flow and elevation data compared to this model. The heavy dashed lines show the model calculated using the parameters listed in (7) and the equation listed in (6). The solid line in the lower part of the figure shows the depth versus age curve of Le Pichon, Francheteau and Bonnin (their "fast" curve shown in Figure 4); as is to be expected, the theoretical curve fits this curve very well. In the top of Figure 5, the "dots" are the average heat flow points of Sclater and Francheteau (1970), and the two wavy solid lines are the "50%" lines of Figure 3. (The light dashed line is the theoretical fit of McKenzie). The heat flow predicted by the model here is much higher than the observed points in a region from the crest out to about an age of 30 MY. This is even more apparent in Figure 3: the dashed lines drawn near the top of these diagrams show the heat flow predicted by this model. In fact, it appears as though no data points are above

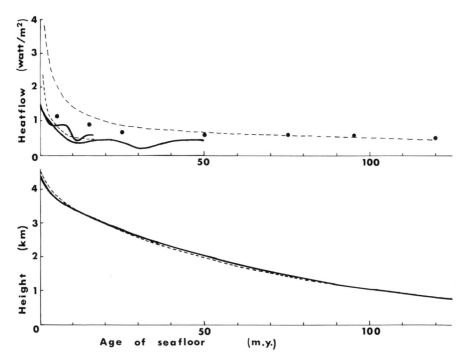

Figure 5. The dashed lines show the predicted heat flow and height of sea floor versus age for the model presented here. (The light dashed line in the top part shows the heat flow predicted by McKenzie as shown in Figure 3.) The solid lines and heavy dots show data: the solid line in the bottom half is the empirical curve of depth versus age (fast) of Le Pichon, Francheteau, and Bonnin, the solid lines in the top half are the "50% lines" shown in Figure 3, and the heavy circles are the average heat flow versus age points used by Sclater and Francheteau (1970). The model here fits the depth versus age data and disregards the heat flow results.

this theoretical curve—the curve acts as an upper envelope of the scattered observed values. If we are to believe that the gradual sinking of the ocean floor away from rise crest is produced by heat loss of the lithosphere, we are forced to conclude that much more heat is lost from the ocean floor than is measured by heat flow probes.

A heat flow measurement consists of measuring the thermal gradient in the sediment of the ocean floor, measuring (or estimating) the thermal conductivity of the sediment, and multiplying these together to get the heat flux conducted through sediments. The magnitudes of the effects of the ice age, sedimentation rates, vertical migration of water through the sediments as the deeper sediment layers below are compacted, and many other effects have been estimated and are considered negligible. The large scatter in the measurements and the apparent bias toward lower than expected values near rise crests has long been a problem in heat flow interpretation, and three mechanisms have been advanced to explain this effect. The first two capitalize on the bias in areal distribution of successful heat flow measurements: if the probe falls on bare rock, no measurement will be made; if it falls on a sediment pond, the measurement may be successful. The first mechanism utilizes the fact that small sediment ponds will have a lower heatflow than bare bedrock nearby. The blanketing effect of the poorer conducting sediments deflects some heat to neighboring areas, hence in areas of rough topography the successful measurements will be biased to the low side. However, to be really effective in deflecting the heat, the width of the sediment ponds should not be significantly wider than they are deep, and very few areas have the topography required to produce significant anomalies by this mechanism. (If the sediment cover is continuous and there are no bare spots, then the topographic effect can produce some scatter in the measure value but not a bias toward low values.) The second mechanism utilizes the fact that slumping is more likely to have occurred into a low sediment pocket than on a higher peak of bare basement. Since slumping mixes up stratifications and hence flattens the thermal gradient, an area into which a slump has occurred will have a low heat flow until the appropriate thermal gradient is re-established. The difficulty with this model is that at each site of low heat flow, we must assume that the basin has been filled with a few meters of sediment within the last few years—a highly improbable coincidence. The third mechanism, which now has widespread support, is that the large scatter and bias toward low measurements results from appreciable heat transport by thermal springs extending deep into the crust.

Such thermal springs were proposed by Elder (1965), based on an analogy with subaerial regions of high heat flow. Geochemical arguments have been advanced (Deffeyes, 1970; Bottinga and Allegre, 1973) to estimate the amount of seawater circulated in cracks deep in the ocean crust. Deffeyes has shown that the circulation needed to effect an ion exchange of Na, Mg, K, and Ca between seawater and basalt of the amount needed to transform the basalt of layer 2 into metabasalt is of the magnitude needed to transport most of the heat flow out of the newly formed ocean floor. In Deffeyes's model, near the rise crest there would be many hot springs spaced a few kilometers apart with seawater circulating several kilometers deep in cracks a few millimeters wide. The thermal springs would die out farther from the rise crest—either because the cracks fill up with deposits or because continued tectonic activity such as is found only near the rise crest is needed to keep the cracks open. In particular, such cracks might grow near the crest because of rapid thermal contraction due to cooling. With this model, there could be numerous localities of hydrothermal concentration of minerals scattered all over the ocean floor, concentrations analogous to those on Cyprus. Active hydrothermal activity would occur only near rise crests, but its effects would be preserved on all parts of the ocean floor.

Direct evidence of such thermal springs has been obtained at one location on the Galapagos Rise (Williams et al., 1973). Near the equator there is a high sedimentation rate, and there is a sufficiently thick sediment cover in which to insert a heat flow probe very close to the rise crest. Seventy heat flow measurements were made in a very small area, and the spacing of the measurements was sufficiently close so that adjacent stations gave values highly correlated with each other. Most of the measurements were very low, but there were some groups of readings that were very high—in one group up to 1200 mw/m^2 (30 HFU). The most positive demonstration that hot springs transport most of the heat flow at rise crests is still to be made, i.e. to find very hot water or evidence of hydrothermal deposits in the sediment at the site of such a very high heat flow measurement. However, in the absence of such conclusive evidence, we can still state that the hot spring mechanism appears to relate satisfactorily a variety of observations: (1) the localized pockets of very high heat flow observed by Williams et al., (2) the large scatter in the many heat flow measurements made near rise crests and the lesser scatter in measurements made far from the crests, (3) the low (compared to theoretical models) average heat flow at rise crests, (4) the evidence of hydrothermal mineralization of some parts of the seafloor as found in some DSDP holes and by analogy with Cyprus, (5) the observation of sediments near rise crests enriched in some minerals (Bostrom and Peterson, 1966), and (6) the exchange of Na $\leftrightarrows$ K, Mg $\leftrightarrows$ Ca between seawater and ocean floor needed to explain the observed ion concentration of seawater.

The hot spring model produces a bias toward low measurements of heat flow (3 above) in the following manner. Where water is descending into the crust, or circulating horizontally deep beneath the surface, it cools the material below and reduces the thermal gradient and hence reduces the heat flow measured at the surface. Where the water rises, the outward heat flux is higher than normal. The conservation of energy requires that the excess of heat carried out at the upwelling region equals the deficit of heat in the downwelling and horizontal flow regions; however two factors are present to bias the measurements. First, the upwelling areas appear to be very small, thus the large fraction of measurements will be made in the "lower" areas. If the high heat flow areas are very small and very high (some places on land have heat flows 1000 times that of neighboring areas), with a finite number of samples the average will be biased toward a lower than true average. The second bias is due to the design of the heat flow instrument. Present instruments operate on the assumption that heat is conducted through the sediments and that the thermal conductivity and temperature gradient in the sediment are the only quantities which need be determined. The first bias could be eliminated by making many closely spaced measurements. The second bias could be removed only by constructing a new type of heat flow device that would measure porosity, water temperature, pressure gradients, and other parameters that would determine the amount of heat convected through the sediments.

Before returning to our main problem of how heat flow relates to vertical motion in the crust, we would like to digress and examine how water circulation might affect another feature of the observed heatflow pattern—the small minimum in average heat flow that is observed some distance from the rise crest (see for example Figure 3). The heat flow versus age of the ocean floor patterns in Figures 1 and 5 were computed assuming that the vertical temperature profile along the crest of the rifting plates, i.e. the temperature at $x = t = 0$, was a constant T_0 from top to bottom (or a minor modification, that the vertical temperature gradient on the axis followed the adiabat). If we somehow cool the material to great depths at the axis, then this boundary condition of our model is changed and the a_n of equation (5) will have different values and the heat flow versus age will be different. Figure 6 shows the heat flow patterns that would result from several patterns of initial temperature along the axis. Thus if water could circulate in cracks 10 km deep, and at the rise crest the tensional stresses of spreading could possibly permit this, then, as the figure shows, there could be a small heat flow minimum next to the crest. (Implicit in this model is the requirement that such deep cracks close as soon as the newly created floor

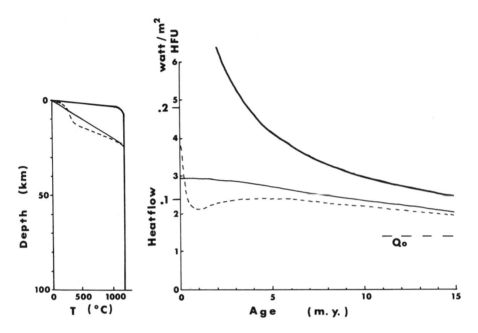

Figure 6. The heat flow versus age pattern can be dramatically changed if the initial temperature pattern at $x = 0$ can be changed. The heavy solid line shows the heat flow if the temperature at the origin is hot up to very close to the top surface, the light solid line shows that the heat flow anomaly at the crest largely disappears if the crust at the crest can be cooled to 20 km depth, and the dashed line shows that a heat flow minimum next to the rise axis can occur if there is very abrupt cooling to 10 km depth at the crest.

leaves the crest so that the heat is *conducted* up to the surface.) This might be a possible (though unlikely) mechanism to produce the observed minimum, but we prefer another hypothesis. Suppose there are numerous small shallow thermal springs and that they transport a sizable fraction of the heat flow for several million years after the formation of the floor. Heat flow measurements will give lower than true heatflux values because the measurements see only the conductive part of the total. Then suppose that the numerous small springs close up. The thermal gradient of the uppermost crust and the sediments will increase, with conduction alone transporting out the heat formerly moved by conduction and convection; thus the measured heat flow will rise and mark the outside limit of the minimum.

We would now like to examine the consequences of the model presented in (7) and (8) and Figure 5. The primary feature of this model is the large thickness of the cooling slab, $H = 150$ km. This is thicker than other models because it was obtained by fitting the reliable depth-age data without compromising the parameter to partially fit the less reliable heat flow-age data. The lithosphere is widely regarded as having zero thickness at a rise crest and then becoming thicker with increasing age to a maximum thickness of about 70 km. (Different seismic measurements of lithosphere thickness, or the top of the low velocity zone, give different values, but the range 50 to 100 km fits all of the data.) In this model, the cooling slab is much thicker than just the lithosphere; both asthenosphere and lithosphere cool as they move away from the spreading center.

Temperature versus depth profiles for various ages of floor are shown in Figure 7 against a background of lines that represent rheological divisions in the mantle. From top to bottom, these lines are estimates of the solidus, percent partial melt, liquidus, and a division of the solid mantle into regions of high creep rate and negligible creep rate. The division of the upper mantle into two distinct categories—lithosphere with negligible creep and asthenosphere with appreciable creep—is surely a gross oversimplification. Nevertheless, the main features of the mechanical properties appear to be represented by this division and it is conceptually simpler to consider just these two categories. We assume that the boundary between negligible creep and appreciable creep occurs at 0.75 of the melting temperature of the mantle (in °K); this is shown on Figure 7 as the boundary between the unstippled area at the bottom (lithosphere) and the area of light stippling (asthenosphere).

The solid lines show temperature profiles for various ages as calculated with formula (6) and parameters (7). The "pivoting" of the curves at $T = 1325°C$ and $z = 150$ km (the result of our boundary condition that the base of the slab remain at a constant temperature) looks artificial; is this an artifact or is it necessary to match the observed surface features? To investigate this, we solved equation (1) for the infinitely thick slab, or half-space. In this case the solutions involve the error

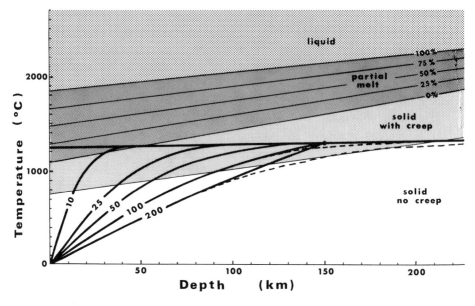

Figure 7. Temperature versus depth profiles for ocean floor of different ages (in MY). The solid lines were computed using the "slab" formulas given in (6) and the parameters listed in (7). The dashed lines (coincident with the solid lines for ages 10, 25, and 50 MY) were computed using the "half-space" formula given in (9) with the parameters listed in (7). In addition to the temperature profiles, lines specifying mechanical properties of the mantle are shown; the solidus, liquidus, and several percent partial melt lines and a line (0.75 of the melting temperature) which we use to divide the solid mantle into two categories—a part with appreciable creep and a part with negligible creep. Where the temperature profile intersects this line marks the depth of the asthenosphere/lithosphere boundary. Note the "half-space" model for 200 MY would have cooled so much there would be no asthenosphere.

function erf $(x) = \dfrac{2}{\sqrt{\pi}} \displaystyle\int_0^x e^{-y^2}\, dy$ and its derivatives and integrals instead of the fourier series. In this case we get

$$T(t,z) = T_0\, \mathrm{erf}\left(\left(\frac{\rho c}{4Kt}\right)^{1/2} z\right) + \gamma z$$

$$Q(t) = \left(\frac{KT_0\rho c}{\pi}\right)^{1/2} \frac{1}{t^{1/2}} + K\gamma \qquad (9)$$

$$h(t) = h_0 + \left[\frac{\rho_m}{\rho_m - \rho_w} 2\alpha T_0 \left(\frac{K}{\pi\rho c}\right)^{1/2}\right] t^{1/2}$$

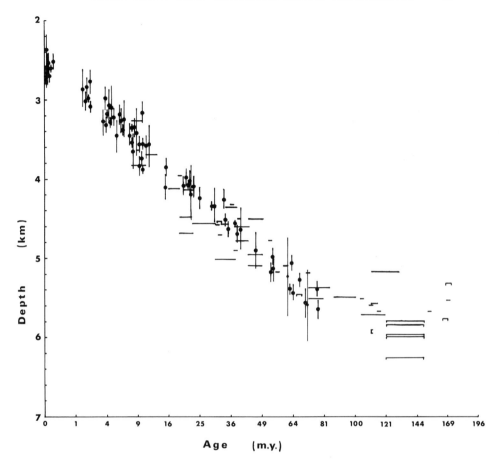

Figure 8. The depth of the ocean floor plotted versus the square root of the age of the floor. The circles with vertical lines show the average depth and standard deviation of typically 5–10 locations of the same age in a given ocean as tabulated by Sclater, Anderson, and Bell (1971). The horizontal lines show the depth of basement of a JOIDES hole; the horizontal extent of each line shows the uncertainty in age of the hole. (From Sclater and Detrick, 1973.) The symbols ⊓ and ⌐ indicate that basement was not reached or the site was older, respectively. (The depths are corrected for sediment thickness as described in the text.) The straight line fit predicted by formula (9) breaks down at about 90 MY, indicating that the "half-space" feels the effect of the bottom of the slab at this age.

Terms for a constant initial thermal gradient γ have been included, as they were for equation (6).

Acknowledgments

A large part of this paper was written while I was at the Centre Oceanologique de Bretagne, Brest, France. I thank many persons at C.O.B. for their help, especially Jean Francheteau. This work was partially supported by the National Science Foundation, grant GA 31364.

References

Anderson, R. N., and E. E. Davis, 1973. A topographic interpretation of the Mathematician Ridge, Clipperton Ridge, East Pacific Rise system, *Nature,* v. 241, pp. 191–193.

Anderson, R. N., and D. P. McKenzie, and J. G. Sclater, in press. Gravity, bathymetry and convection in the earth, *Geophys. J. Roy. Astr. Soc.*

Bostrom, K., and M. N. A. Peterson, 1966. Precipitates from hydrothermal exhalations on the East Pacific Rise, *Economic Geology,* v. 61, 1258–1265.

Bottinga, Y., and C. J. Allegre, 1973. Thermal aspects of sea-floor spreading and the nature of the oceanic crust, *Tectonophysics,* v. 18, pp. 1–17.

Deffeyes, K. S., 1970. The axial valley: A steady-state feature of the terrain, in *Megatectonics of the Continents of Oceans,* ed. by H. Johnson and B. L. Smith, Rutgers Univ. Press, New Brunswick.

Elder, J. W., 1965. Physical processes in geothermal areas, in *Terrestrial Heat Flow,* ed. by W. H. K. Lee, Geophysical Monograph Series 8, Amer. Geophys. Union: Washington.

Ito, K., and G. C. Kennedy, 1967. Melting and phase relations in a natural peridotite to 40 kilobars, *Amer. J. Sci.,* v. 265, pp. 519–538.

Le Pichon, X., J. Francheteaun, and J. Bonnin, 1973. *Plate Tectonics,* Elsevier, Amsterdam.

McKenzie, D. P., 1967. Some remarks on heat flow and gravity anomalies, *J. Geophys. Res.,* v. 72, pp. 6261–6273.

Menard, H. W., 1969. Elevation and subsidence of oceanic crust, *Earth Planetary Sci. Letters,* v. 6, pp. 275–284.

Parker, R. L., and D. W. Oldenburg, 1973. Thermal model of ocean ridges, *Nature,* v. 242, pp. 137–139.

Sclater, J. G., R. N. Anderson, and M. L. Bell, 1971. The elevation of ridges and the evolution of the central Eastern Pacific, *J. Geophys. Res.,* v. 76, pp. 7888–7915.

Sclater, J. G., and R. Detrick, 1973. Elevation of midocean ridges and the basement age of JOIDES deep sea drilling sites, *Geol. Soc. Amer. Bull.,* v. 84, pp. 1547–1554.

Sclater, J. G., and J. Francheteau, 1970. The implications of terrestrial heat flow observations on current tectonic and geochemical models of the crust and upper mantle of the earth, *Geophys. J. Roy. Astr. Soc.,* v. 20, pp. 509–542.

Schubert, G., and D. L. Turcotte, 1972. One-dimensional model of shallow-mantle convection, *J. Geophys. Res.,* v. 77, pp. 945–951.

Williams, D. L., R. P. Von Herzen, J. G. Sclater, and R. N. Anderson, 1973. Lithospheric cooling on the Galapagos Spreading Center (GSC) East Pacific, *Trans. Amer. Geophys. Union,* v. 54, p. 244, (abstract).

Wyllie, P. J., 1971. Role of water in magma generation and initiation of diapiric uprise in the mantle, *J. Geophys. Res.,* v. 76, pp. 1328–1338.

The earth would have no fossil fuels were there not sedimentary basins that serve as sites of petroleum generation, accumulation, and storage. These basins are of diverse sorts, with widely different kinematic histories, different types of sediments, and different amounts and styles of petroleum occurrence. Fischer, in the following paper, assesses the various mechanisms that may be invoked to explain the development of such basins. This development is in all cases a complex one, in which one factor engenders or triggers others.

Origin and Growth of Basins

ALFRED G. FISCHER[1]

ABSTRACT

Basins in the geological sense are depressions in the basement surface (base of the layered rocks) relative to sea level. They may be *stuffed* (largely filled with sedimentary or volcanic matter) or *starved* (largely filled with hydrosphere or atmosphere).

Primary basins are floored by oceanic lithosphere, which has a limited existence span (less than 200 MY). They include normal oceanic basins, platillo basins, interarc basins, and possibly additional types.

Secondary basins result from geological modifications of primary basins or of continental platforms. The majority develop under isostatic conditions. While it is conceivable that some volcanic basins involve only a transfer of magma from deeper layers in the crust to the surface, with little change in buoyancy, the majority of isostatic basins result from decrease in lithospheric buoyancy. This can follow from any of a number of geological modifications of the lithosphere, such as shrinkage on cooling, tectonic attenuation, loading, or phase changes.

[1] Department of Geological and Geophysical Sciences, Princeton University.

Alternatively, subsidence of lithosphere may occur in response to development of asthenospheric inhomogeneities. Finally, it appears likely that isostasy is violated in very mobile areas such as subduction zones, where trenches are dynamically maintained for limited spans of geological time. The history of basins in general involves the interplay of several contributory causes. Loading is a major secondary factor in most. A survey of the temporal subsidence patterns of various basin types suggests that a wide gap remains between general theory and specific understanding. Plausible explanations are now at hand for the development of the primary and secondary oceanic basins, for rift basins in continental crust, and for the basins developed along trailing continental margins (see the following paper by Kinsman). On the other hand, the basins of the stable cratonic interiors, as well as the synorogenic and postorogenic basins of mobile belts, have not as yet found specific and coherent historical explanations.

Introduction

One of the fundamental characteristics of the earth is the mobility of the lithosphere: expressed on the one hand by lateral motions now reckoned by the more mobilistic in thousands of kilometers; and on the other in vertical displacements of kilometers and tens of kilometers. The result is a topographic relief of near 20 km, and a tectonic relief that is very much larger.

To the topographer or geographer, the earth's basins are 3-dimensional depressions of the lithospheric surface, filled with water or with air. To the geologist, they take on the fourth dimension of time, and include the layered rocks formed at the earth's surface—the sediments and volcanics that may accumulate to thicknesses of many kilometers. To him, the shape and depth of the basin are not defined by the land surface or the sea floor, but by the deeper surface of a geological basement of plutonic or metamorphic rock complexes. The filling of sediments and volcanics records the history and development of the basin. In terms of their fill, basins form a spectrum from *stuffed* ones, essentially brim-full of sedimentary or volcanic matter, to *starved* ones (Adams and Frenzel, 1951), largely filled with water. A given basin may, of course, change from one to another in time or space: a classical example of this is the Adriatic-Po basin, the northwestern end of which (Po Valley) is stuffed with sediments poured off the Alpine welt, while the mid-Adriatic shows appreciable water depths, and the southern Adriatic, showing water depths of several kilometers, is starved.

The layered rocks filling basins include much of the world's mineral resources, inculding all of its petroleum. Among authors who have dealt extensively with various aspects of basins, we note in particular Bucher (1933), Kay (1951), Beloussov (1962), Wyllie (1971), and Hsü (1965). Halbouty et al. (1970) provided a general classification of sedimentary basins, reviewed by Klemme in this volume. Dallmus (1958) dealt with the deformational aspects of basins subsiding on a spherical earth.

The present paper, written within the general framework of plate tectonics, attempts first to assay the causes of basin development. It then goes on to a survey of basin kinematics, comparing the subsidence patterns of several distinct types of basins. The origin and development of a particular type of basin—that formed on "linked" or Atlantic-type continental margins—is touched on here, but is more extensively developed in the succeeding paper by Kinsman.

Ideas expressed here have been discussed in particular with my colleagues K. Deffeyes, D. J. J. Kinsman, W. J. Morgan, and R. Phinney, for whose help I am grateful, but who are not to be held responsible for the concepts or their manner of presentation.

FUNDAMENTAL ASSUMPTIONS

The fundamental assumptions underlying the following treatment of basins are mainly those of plate tectonic theory (Morgan, 1968; Oxburgh, 1971). The outer portion of the solid earth (Figs. 1, 4) is con-

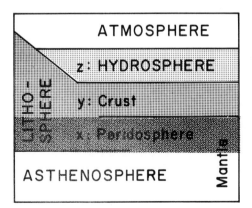

Figure 1. Components of isostatic system.

sidered to have a comparatively brittle shell, the *lithosphere* (Daly, 1940; it is the tectonosphere of Morgan, 1968), generally some 100–200 km thick. This includes both the basaltic-granitic crust, and the uppermost brittle portion of the presumably ultramafic mantle which Mac-Donald (1972) has termed the *peridosphere*.

This lithosphere rests on a substrate—belonging to the mantle—that is characterized by higher temperatures and plastic behavior and is termed the *asthenosphere*. The contact is probably gradational, at temperatures in the range of 800 to 1000°C (Morgan, this volume, Fig. 7), and corresponds to the seismic low-velocity zone.

The lithospheric shell is viewed as broken into a limited number of plates, moving about in a somewhat independent manner. Where they diverge, asthenospheric material wells up between them. Differentiating and cooling, it forms the juvenile (oceanic) lithosphere. Along other margins, the lithospheric plates shear past each other, while yet elsewhere one plate plunges into the asthenosphere—a process which Ampferer and Hammer (1911) termed "swallowing" (*Verschluckung*), and which is now more widely known as *subduction*.

For a model of lithospheric structure, in terms of rock types and densities, we shall use that of Holmes, 1965 (Fig. 2; see also Kinsman, Fig. 1). Note that the representative columns shown are only 60 km

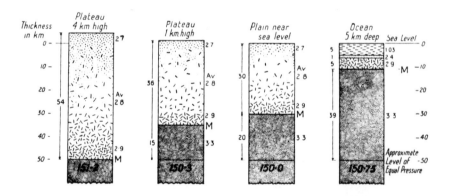

Figure 2. Columns of equal cross section through characteristic parts of the continents and ocean floor. The figures to the right of the columns indicate approximate densities. These, multiplied by the appropriate thicknesses, down to a depth of 50 km, give a total figure proportional to the pressure at that depth. These columns represent only about the upper half of the normal lithosphere. Modified from Holmes, 1965, with permission of Thomas Nelson and Sons, Ltd., and Ronald Press Co.

deep, representing perhaps the upper third to one half of the lithosphere postulated here.

For practical purposes, we may consider any lithospheric column as composed of a finite number of superimposed layers (Fig. 1; see also MacDonald, 1972), each characterized by a specific density. The lowest layer x (peridosphere) is essentially frozen mantle. Overlying layers of the lithosphere, of lesser density, are $a, b, \ldots, n$, collectively constituting the crust, y. This crust is, in most cases, overlain by a layer of water, z. Under isostatic conditions, P, the pressure at some reference depth in or at the base of x (Holmes used 50 km, while Kinsman, this volume, uses 100 km) is the sum of the masses of $x + y + z$, and is the same for all columns, namely about 3.15×10^7 g/cm^2 at 100 km.

In certain belts these models and the principle of isostasy do not apply: In zones of lithospheric growth, such as the axis of the mid-oceanic ridge system, there is essentially no lithosphere. Along subduction zones, the asthenosphere becomes complicated by the introduction of lithospheric slabs—cooler, brittle, and containing basaltic and sialic matter. Furthermore, there is good reason to believe that the trenches associated with subduction belts are areas of deficient mass, far out of isostatic balance, and maintained by the dynamics of subduction—a matter further discussed below.

While there is a wide spectrum of lithospheric structure of which the columns in Figure 2 are inferred examples, two kinds of lithosphere are vastly more abundant than all others put together (Bucher, 1933): the *oceanic lithosphere* characterized by a thin mafic crust (the 4th column in Fig. 2), and the normal *continental lithosphere,* characterized by a thick crust of largely granitic composition (the 3rd column in Fig. 2). These frequency maxima of lithospheric types find topographic expression in the two dominant levels of the lithospheric surface: the ocean floors and the broad continental platforms (Fig. 3). Plate tectonics has given us an explanation of the origin of these surfaces; the oceanic lithosphere is juvenile lithosphere, most of which is created in the gap between rifting plates (left side, Fig. 4): The continental lithosphere (Figs. 3, 4) results from the wholesale reordering of lithospheric material in mobile subduction zones (center, Fig. 4). The process probably involves a number of separate phenomena, among them the "obduction" transfer of sediment from the surface of the downplunging plate to the lip of the overriding plate margin; the injection of intermediate and acidic plutons and volcanics from below; the loss of dense (ultramafic) matter

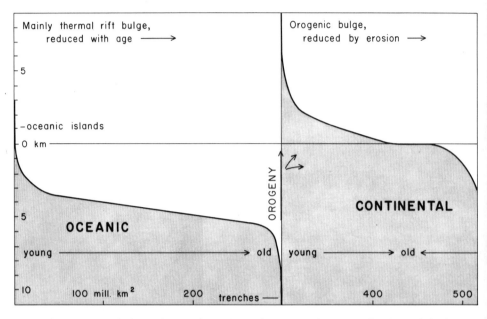

Figure 3. Evolution of oceanic and continental surfaces. A simple statistical (cumulative) frequency plot of the lithospheric surface—the well-known hypsographic curve of Kossina (1921) (Bucher, 1933; Holmes, 1965)—clearly demonstrates the existence of two dominant topographic levels—the sea floor and the continental platform—which correspond to distinct types of lithosphere. However, the elevations between about −1 and −3 km contain an appreciable admixture of two basically different types of lithosphere: thinned continental lithosphere and thermally expanded oceanic lithosphere (rift bulge). Figure 3 is an attempt to separate these, in providing separate frequency plots for oceanic and continental lithosphere. The shoaler parts of the oceanic lithosphere are due in part to excess volcanism (platillo crust), in part to the thermal bulge associated with rifting. The trenches are probably anisostatic downwarps. The oceanic lithosphere tends with age toward depths of −6 km. The higher parts of the continental lithosphere are underlain by orogenically over-thickened crust, which erosion tends to reduce to sea level; the parts of continental crust that are immersed below the hydrosphere are either not yet sufficiently thickened by orogeny (MacDonald's tectonitic crust) or have been thinned by secondary geological processes (tectonic attenuation, truncation of thermal bulges, etc.). They tend to be thickened with time, either by orogeny or by sedimentation. Thus the continental lithospheric surface tends to evolve toward sea level.

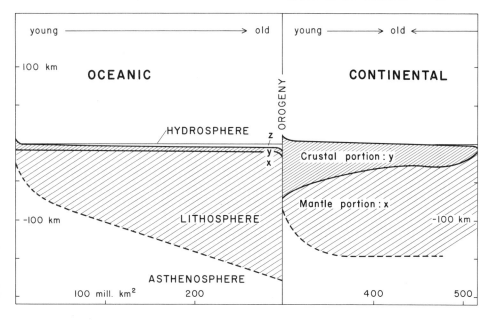

Figure 4. Evolution of lithospheric structure. The juvenile (oceanic) lithosphere is visualized as thin at birth, and as gradually thickened by accretion of x (peridosphere) at its base. y, the crust, tends to grow slightly, by sedimentation; z, the hydrospheric cover, increases as the lithosphere ages, cools, and subsides. Orogeny, including subduction and a complex of related processes, thickens and differentiates the crust and disturbs the simple lithosphere-asthenosphere relationship. High associated heat flow probably raises the asthenosphere-lithosphere boundary, but with age x will presumably again accrete at the base to restore lithosphere thickness. Excessively thick sialic crust rises to form mountain ranges, but surficial erosion tends to reduce elevation to sea level and crustal thickness to 30 km. Beyond this point, tectonic or other geological processes may thin the lithosphere and cause it to submerge, but sedimentation and orogenic processes will tend to restore it.

from the base of the overriding plate; and lateral compression of the overriding plate, leading to a corresponding thickening.

Whatever the complex processes which are involved in these mobile belts, and which we can ignorantly but conveniently sweep together in the general concept of orogeny, they tend to thicken y, the crustal part of the lithosphere, which may grow to 60 km or more (center of Fig. 3). It is thereby endowed with a buoyancy that, with return of isostatic conditions, lifts its top out of the sea, to form mountains (hence the term *orogeny*). The surplus of such an excessively continental lithosphere is subsequently removed by the interplay between erosion and renewed uplift, until an equilibrium position is attained, at which the lithospheric surface remains near erosional base level—for practical purposes, sea level (3rd column of Fig. 2). This normally occurs when *y* has been reduced to a thickness of about 30 km (toward right side, Fig. 4).

Thus, of the two main types of lithosphere, one forms the great oceans—the *primary basins* of first order—while the other, created in mountain belts subsequently reduced to peneplains, forms the *continental platforms*. A more detailed breakdown of lithospheric or crustal types, by MacDonald (1972) recognized five types: The *normal oceanic crust;* a variant of this with excessive thickness of basalt, termed *platillo crust;* a *transitional crust* of small ocean basins; a *tectonitic crust* which is a young, thin continental crust (right end, Fig 4); and the *continental crust sensu stricto,* characterized by extensive granitic-metamorphic shield terranes.

Whatever classification of crusts or lithospheres one may choose, the fundamental types of lithosphere, underlying the primary ocean basins and the continental platforms, continue to be modified by geological processes. These result in areas of uplift and areas of subsidence, and are thus the causes of secondary basins.

Primary (Oceanic) Basins

A first-order division of the earth's surface recognizes *oceanic basins* and *continental platforms,* each corresponding to a different type of underlying lithosphere (Figs. 3, 4). The continental lithosphere, having a mean density smaller than that of the asthenosphere, tends to float on the latter indefinitely, and contains within itself a record of nearly all of earth history. The oceanic lithosphere, forming the primary basins of the earth, has markedly different properties. It has a mean composition and therefore a mean density not very different from that of the asthenosphere. When freshly formed and hot, as along the midocean ridge system, it is probably somewhat less dense as a whole. As it ages and cools (see Morgan, this volume) it presumably becomes denser than the plastic asthenosphere, on which it rests in an unstable state, maintained due only to the viscosity of the asthenosphere. Any buckling of the lithosphere, as in a plate collision, will trigger its foundering into the depths. With great age it may become dense enough to founder spontaneously—the process that Moberly (1972) has evoked to explain the marginal or interarc basins of the western Pacific. To judge from the presence of mid-Jurassic sea floor in the western Atlantic (Hollister and Ewing, 1972), which has not as yet foundered, and from the inferred Jurassic or Triassic age of the foundering sea floor in front of the Japanese, Bonin, and Mariana Arcs (Fischer, Heezen et al., 1971), the critical age for such spontaneous foundering appears to be on the order of 170–200 MY.

Thus the primary basins differ from all others in several important respects: they are characterized by juvenile lithosphere, which has a limited life-expectancy; many and perhaps all of them grow laterally; they also vanish laterally, i.e. at some margin or other. They are the largest basins of all, by an order of magnitude, and are almost entirely starved—being many times as large as any possible sources of fill.

The primary basins include at least three subtypes: *normal oceanic basins, aseismic ridge* or *platillo basins,* and *interarc basins.*

Normal oceanic basins are underlain by lithosphere grown in a normal midoceanic ridge setting, at a mean water depth of some 2.5 km, and seem to be characterized by perhaps 1.5 km of basalt over the hypabyssal "layer 3" of seismologists.

A different type of lithosphere is created at centers of intense volcanic action, such as Iceland or the Azores—emergent areas that are the generating points for the aseismic ridges, running transversely across the main ocean basins. These centers of generally more alkalic vulcanicity appear to build an excessively thick upper basalt layer, resulting in a more-than-normally buoyant, oceanic lithosphere, which therefore rises above the normal ocean floor. This is MacDonald's (1972) *platillo crust.* Topographically this lithosphere stands high above the normal ocean floor. But if we define the basin concept geologically, as limited by the base of the layered rocks, then we must define our basins by the base of the layered basalts, and this may well lie deeper in platillo crust than under the normal sea floor. Thus, paradoxically, the topographic aseismic ridges may actually be geological basins relative to normal oceanic areas, but this matter remains academic until drilling has provided a better insight into the nature of the igneous sea-floor complexes.

The third type of oceanic basin is that developed behind island arcs (Karig, 1971, 1972; Fischer, Heezen, et al., 1971; Moberly, 1972). This appears to have somewhat different seismic properties, to lack well-developed magnetic stripe anomalies, and to lack a well-defined generating ridge. Its manner of growth remains largely mysterious.

Perhaps the Black Sea, the Caspian, and other isolated small seas represent yet other types of oceanic lithosphere, and other subtypes of primary basins.

SECONDARY BASINS

As *secondary* we may characterize those basins that are formed by secondary modifications of primary oceanic basins, or of continental platforms. Such modifications would seem to arise in four major ways, as follows (Fig. 5):

I. Modifications in an isostatic regime.
   A. Modification of lithosphere by simple rearrangement, involving little change in buoyancy.
   B. Loss of lithospheric buoyancy,
      1. by local modification of the lithosphere—such as volume changes (due to thermal effects or phase changes), tectonic or erosional thinning, loading, or injection of dense matter.
      2. by development of asthenospheric inhomogeneities.
II. Local development of anisostatic downwarps—probably mainly in association with subduction (trenches).

We shall briefly examine each of these, in turn.

Basin Formation by Redistribution of Matter within the Lithosphere

It seems reasonable to suppose that some forms of volcanic activity involve the collapse of magma chambers within the lithosphere, accompanied by a corresponding sag in the overlying basement surface, and the filling of this basin with the volcanic ejecta and interbedded sediments (Fig. 5E). Such events need not involve major changes in the buoyancy of the lithosphere affected. Perhaps the great volcanic basin fill of the Permian Bozen quartz porphyry, in the southern Alps, is of this origin.

Shrinkage of Lithosphere

If the lithosphere shrinks in volume without change of mass, the result (short of foundering) will be a subsidence of the lithospheric surface,

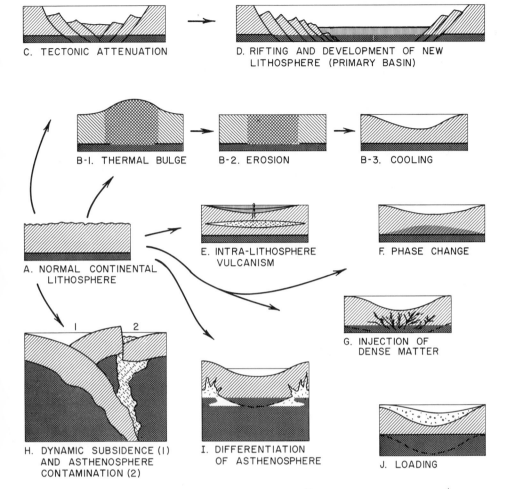

C. TECTONIC ATTENUATION

D. RIFTING AND DEVELOPMENT OF NEW LITHOSPHERE (PRIMARY BASIN)

B-1. THERMAL BULGE

B-2. EROSION

B-3. COOLING

A. NORMAL CONTINENTAL LITHOSPHERE

E. INTRA-LITHOSPHERE VULCANISM

F. PHASE CHANGE

G. INJECTION OF DENSE MATTER

H. DYNAMIC SUBSIDENCE (I) AND ASTHENOSPHERE CONTAMINATION (2)

I. DIFFERENTIATION OF ASTHENOSPHERE

J. LOADING

Figure 5. Possible mechanisms of basin formation. Normal continental lithosphere, riding with its surface near sea level (A) may be induced to sag into basins by a variety of mechanisms. In B, a temporary bulge induced by heating is removed by erosion and becomes the site of a basin when temperatures return to normal. In C, a basin is formed by tectonic attenuation (rifting), which, if continued to the stage shown in D, leads to complete disruption of the continental lithosphere, and to the growth of a new juvenile ocean basin. Growth of a surficial basin at the expense of a deeper magma chamber is shown in E. F depicts the effect of a phase change at depth, G that of injection of dense mantle matter into the crust. H portrays processes associated with subduction zones: at 1, the development of a trench, out of isostatic equilibrium and maintained by the lag between plate subsidence and asthenosphere response; at 2, the development of inhomogeneities in the asthenosphere, by generation of sinkers and rising magmas from the downgoing plate. I portrays the possibility that mantle differentiation leads to generation of magmas under continents, which partly invade the lithosphere and produce uplifts and downwarps (swells and swales). J reminds us that any basin, whatever its cause, will be further depressed by surficial loading by water, sediments, volcanics, or some combination of these. Lithosphere cross-hatched, asthenosphere stippled. Not to scale.

i.e., a basin will develop. Such shrinkage may be caused by two factors: cooling and phase changes at depth.

The best example of basin formation due to shrinkage is provided by the oceans. As pointed out above, juvenile oceanic lithosphere is normally formed in a rift bulge, the midocean ridge system, with a mean elevation of −2.5 km. As the lithosphere ages and moves away from the generating rift, it cools and shrinks, and its surface subsides to form the flanks of the ridge, to reach depths of 5.5 to 6.5 km. (Menard, 1969; Sclater and Francheteau, 1970; Morgan, this volume). The rate of subsidence decreases with time, from about 90 Bubnoff units (m/MY, Fischer, 1969) in the first ten million years, to 33 Bubnoffs in the next 30 million years, and to 20 or less Bubnoffs thereafter (Fig. 8; also Kinsman, Fig. 2, this volume; and Morgan, Fig. 8, this volume). The change affects both normal oceanic and platillo basins, and thus overprints the primary basins with a secondary effect that provides paired mirror-image subbasins on either side of the generating rift. Nearly one quarter of the total subsidence is due to water load (to be discussed below) which means that the remaining ±3 km of subsidence are a direct measure of lithospheric shrinkage—amounting to 3% of a 100-km lithosphere or 1.5% of a 200-km lithosphere. It remains uncertain whether this is due entirely to temperature changes, or whether phase changes are also involved, as Oxburgh suggests (1971).

The decay of heat bulges is not limited to the oceans but also occurs in continental lithosphere. Here matters grow more complicated. If a normal continental platform area is heated, as along a juvenile rift or around a center of volcanic activity, it will presumably expand, to form a welt (Fig. 5-B-1). The coincidence of rift grabens with uplifted welts has long been recognized: famous examples are the Vosges and Black Forest uplifts along the flanks of the Rhine Graben, and the Arabian and Ethiopian mountains along the Red Sea rift (Fig. 6). Only the inferred relationship has shifted: whereas Cloos (1939) and others saw the uplifts as the primary feature, and the rift as a secondary one, we now view the rift as a feature of lithospheric magnitude, and the marginal uplifts as a secondarily induced response to the heat flow along the rift. In any case, these uplifts—unlike the midoceanic ridge—are subject to erosion (Fig. 5-B-2), and isostasy will renew uplift and thereby provide for further bevelling: The end result is inevitably a thinning of y along such areas of higher heat flow. Cessation of high heat flow and a return to a normal temperature regime will then produce lithospheric shrinkage

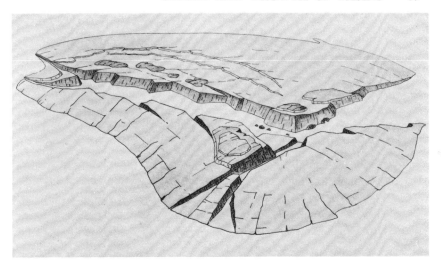

Figure 6. Hans Cloos's drawing of the rift bulge bordering the Red Sea and the Gulf of Aden, from Cloos, 1951, with permission of R. Piper Verlag, Munich.

in x and y, and must induce subsidence (Fig 5-B-3). Unthinned (30-km) crust will return to sea level, while thinned crust will become submerged to varying depths depending on the degree of thinning, as shown in Kinsman's Figures 2 and 3. As Kinsman explains in more detail in the succeeding chapter, this process appears to be the chief initiating cause for basins of the Atlantic margin type, which then continue to undergo depression by loading.

By the same token, one should expect that areas of volcanic activity unrelated to rifting should pass from a hot domed-up stage into a cool subsiding one, and one is led to wonder whether such intracratonic basins as the Illinois and Michigan basins owe their existence to Late Precambrian-early Cambrian centers of vulcanism.

Phase changes in the deeper parts of the lithosphere (Fig. 5-F) are likely to be another important factor in the origin and growth of basins—especially in the mobile belts, where subduction and extensive uplifts move rock bodies through large pressure and temperature ranges. A sizable literature has developed around this subject, and is reviewed by Wyllie (1971). It seems likely that synorogenic basins find part of their explanation here, but the role of these factors in cratonic basins remains wholly obscure.

Tectonic Attenuation of Lithosphere

One of the simplest ways of inducing subsidence is by tectonic attenuation: the stretching and thinning of the crust (y). This reduces P, which is restored by growth of x and of z. The net effect is a depression over the attenuated zone. Such thinning can presumably occur in part by plastic flow ("necking"); but in much of the lithosphere, and certainly in its upper parts, it occurs by normal or "block" faulting, which largely controls the shapes of the basins formed (Fig. 5-C).

The simplest examples of crustal thinning by such faulting are the rift valleys, graben or half-graben basins typically some tens of kilometers wide, hundreds of kilometers long, and several kilometers deep. Large-scale rifting of this sort is represented by the Triassic grabens of eastern North America, the Cenozoic Rifts of Africa, or the Rhine Graben. Such rift valleys are or were accompanied by marginal welts (see above), which constituted a nearby source of sediment; they are therefore generally basins of the stuffed variety.

When rifting along a given belt continues, the attenuation of the lithosphere becomes complete, and a strip of new oceanic lithosphere—a primary basin—appears (Fig. 5-D). The prime example of this in the Red Sea (Lowell et al., this volume). Thus basins of the attenuation type precede the development of primary oceanic basins. Kinsman suggest that disjunct remnants of attenuated continental lithosphere, perhaps 40 to 80 km wide and preceding the development of the Atlantic ocean, lie under the continental slopes and rises of the American and African coasts.

Belts of tectonic attenuation are not limited to rift margins of continents: the Great Basin may be a broad attenuated belt in the continental interior, characterized by a multiple development of half-grabens and an alternation of mountains and basins.

Intense crustal stretching can also occur at convergent plate margins, where the unsupported lip of the overriding plate may be shattered by gravity. This appears to be the case in the coastal basins of northwestern Peru and Ecuador, from the Sechura Desert through the petroliferous La Brea-Parinas and Zorritos areas to the Gulf of Guayaquil (Travis, 1953; Fischer, 1956). Here much of the bedrock has been reduced to a maze of fault prisms at all dimensions from meters to kilometers. Involved are the Paleozoic metamorphic basement and sediments of Cretaceous to Middle Eocene ages; in some areas younger sediments are also strongly affected, in others not. Rising horst blocks are relatively undeformed, whereas in the subsiding grabens the maze of faults has reduced the vertical stratigraphic sequences by at least 30%. Some of the grabens reach depths of 10 km.

"Subcrustal Erosion"

Gilluly (1955, 1964), Alvarez (1972) and others have suggested the possibility that the lithosphere may be thinned from the bottom, by interaction with the asthenosphere. In our model, in which the base of the lithosphere-asthenosphere contact is taken as a thermal and mechanical rather than a chemical boundary, this does not seem to be plausible under stable plates.

Subcrustal erosion and transport of sialic matter into the asthenosphere may occur in orogenic zones, where subduction of lithospheric slabs, generation and movement of magmas, and the growth of thick sialic mountain roots essentially render the simple asthenosphere-lithosphere model inapplicable (Fig. 5-H).

Load Effects

Any load placed upon isostatically balanced lithosphere will raise P, and will be compensated for by subsidence and loss of x (with density 3.3) to the asthenosphere (Fig. 5-J).

The simplest case of loading is perhaps that of a volcanic pile built on subaerial, continental crust. If this pile has a bulk density of 2.5, its load will cause subsidence to the amount of about 82% of its own thickness.

In the case of water and water-laid sediments, we must consider that a given addition of water will induce an additional 33% of subsidence by its own load, or, to put it another way, that water-covered parts of the lithosphere owe 23% of their water depth to the water load itself. Sedimentation on such lithosphere replaces water with sediment, increasing the load accordingly: unconsolidated sediment essentially doubles the load of the water volume it displaces, and therefore doubles the subsidence under it; but, with the accumulation of thicker sediments, compaction sets in, and the density of the sediment increases. This does not cause further loading as such, but makes room above for more water and sediment (and added load). The effects on basin subsidence have been graphed in Kinsman's Figures 6, 7, and 8. Loading by surface additions to y is not confined to sedimentation or volcanism, but also includes slumping and tectonic loading by the emplacement of allochthonous klippen or thrust sheets. The plastic thickening of basin fill, as in fold belts, also adds to y.

An example of the effect of loading is shown in Figure 7, a cumulative subsidence (Bubnoff) diagram of the Michigan basin, in which the progressive subsidence is resolved into an unidentified primary component, and the secondary effects of loading by water and by sediment.

These considerations show that major basins are not readily initiated by loading, but that loading is a contributing factor in any basin initiated by other causes. Its effectiveness rises rapidly with the depth of basin fill: loading of deep basins such as a primary ocean basin, or highly attenuated continental crust, can triple their depths, to reach geosynclinal proportions.

Injection of Dense Matter

Another possible cause of basins is the injection of dense matter—such as ultramafic bodies—into the crust (Fig. 5-G), thereby raising the mean density of y and causing subsidence (reduction of x). Intrusive bodies of ultramafic matter are known from mobile as well as stable shelf regions, but whether such emplacements have ever occurred on a large enough scale to cause the development of a basin remains to be learned.

Variations in the Asthenosphere

So far we have assumed a homogeneous and uniform asthenosphere. This is likely to be an oversimplification.

In orogenic regions, where slabs of lithosphere are plunging into the depths and are carrying with them both temperatures and a chemical composition foreign to those depths, the asthenosphere is certain to develop inhomogeneities ranging from dense sinking bodies to plastic or molten masses of basaltic or sialic composition, which would rise as magmas.

The possibility that the normal asthenosphere also undergoes differentiation, and thus develops inhomogeneities of composition and density, under stable regions cannot be discounted at this time. Furthermore, it seems possible that magmas of intermediate or acidic composition, generated in the asthenosphere or orogenic regions, may become spread under adjacent stable lithospheric segments.

The development of such inhomogeneities in the asthenosphere is sure to initiate a complex of readjusting forces, which will resolve themselves

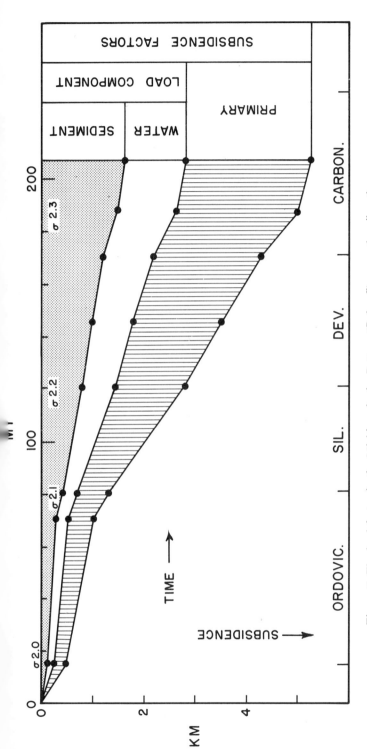

Figure 7. The load factor in the Michigan basin. Bottom (Bubnoff) curve describes the cumulative subsidence of the central parts of the Michigan basin through time, as adduced from the sedimentary fill (see also Figs. 8, 9). This subsidence is a composite response to several factors. A *primary component* results from unidentified factors—unidentified changes in and below the lithosphere. This component is responsible for a basin of limited depth, which underwent further growth as a result of a secondary *load component*, divisible into *water load* and *sediment load*. Drawn so as to provide for a progressive increase in mean sediment density, from 2.1 to 2.3, due to progressive compaction (Kinsman, Fig. 7, this volume). This first approach neglects delays in isostatic adjustments and delays in sediment compaction. Such delays may well account for all of the subsidence in the late stages.

in vertical adjustments. We may imagine the development of an isostatic relief at the asthenosphere-lithosphere boundary, leading to a thickening or thinning of x; a sagging of lithosphere over broader regions in which the upper asthenosphere has been decreased in density; a breakdown of lithosphere into magma chambers developed at its base; or a buoyant rise of lithosphere where its lower portions have been invaded by low-density magmas.

In the cratonic regions, asthenospheric inhomogeneities can be expected to be broad. It seems possible that the "swell-and-swale" pattern (Bucher, 1933) of deformation found here, consisting of "interior basins" separated by "arches," finds its initiation in these processes of asthenosphere differentiation (Fig. 5-I). Such features as the Michigan and Williston basins may be different expressions of this initial cause.

In orogenic belts, the inhomogeneities can be expected to be greater and to be more localized. They provide the most reasonable explanation for the "yoked" developments of adjacent structural troughs and mountain welts (Fig. 5-H-2).

Anisostatic Basins

We have assumed, above, that the oceanic lithosphere becomes gravitationally unstable as it cools and ages, remaining on top of the asthenosphere only because of the latter's viscosity. An isostasy of sorts rules this system despite the instability, up to the line at which the lithosphere finally plunges into the depths: here one might theoretically expect that the combination of a gravitationally descending lithospheric slab with a viscous and imperfectly responding asthenosphere should produce a mass-deficient strip—quite possibly bounded by belts of excess mass. This strip finds its expression in the topographic trenches (Fig. 5-H-1), in filled trench basins, and in the associated negative gravity anomalies. The existence of these trench basins is tied entirely to active subduction: when this ceases, they must rise—and presumably their sedimentary fill becomes deformed in that process, one of the many that we lump under the general term "orogeny."

These basins directly associated with subduction are particularly liable to tectonic emplacement of allochthonous masses, transferred from the submerging plate to the edge of the overriding one (as further developed by Curray, this volume). Furthermore, they are liable to massive emplacement of sedimentary or volcanic matter by slumping—especially when they are superimposed on preexisting starved basins, such as the ocean.

Interrelations of Causes

While we have so far examined independently factors that may cause or contribute to basin formation, it has become abundantly clear that no single factor works by itself: temperature and pressure regimes (and the volume changes related to them) are intimately connected with tectonic, erosional, and sedimentational factors. A minor factor may trigger a major one. Any basin once started, by whatever mechanism, will owe part of its subsequent growth to loading—an effect that generally comes to exceed the initial one (Fig. 7). This may, in turn, trigger phase changes at depth, leading to further subsidence, additional loading, etc. Any attempt to analyze the dynamics of a basin must attempt to separate the various factors (Fig. 7), which not only contributed different proportions to the final basin depth and form, but also operated at different rates, over varying time spans, and in certain sequences. We are only beginning to apprehend the complexities of basin dynamics, and the approach to the genetic understanding of any given basin will have to combine a theoretical evaluation of the various possible causes of subsidence with the kinematic history of the basin as recorded in its fill.

The Widespread Sedimentary Cover over Continental Platforms

Large areas of the continental platforms that are not definable as basins, and that include the tectonic arches between cratonic basins, are covered by a comparatively thin layer—generally less than 1 km—of marine sediment. It seems, then, that the stable continental platforms as a whole show a tendency to undergo subsidence, peripheral to their "old shield" areas, which have tended to remain above sea level.

One cause suggested for such apparent subsidence is an increase in the volume of the hydrosphere through time. Such an increase, leading to an increase in the mean depth of the oceans, would appear at first glance to flood the continental platforms. However, we are not sure how the hydrosphere grew. It seems to be derived from the earth's interior, but whether this outgassing was completed early, occurred at a constant rate through earth history, or followed some intermediate pattern is unresolved. Also, addition of water cannot be simply converted into rise in ocean level, inasmuch as differential loading of the lithosphere and the resulting patterns of subsidence and uplift must be taken into account. Another factor which may account for much of the burial of continental basement surfaces under sediments is a gradual shrinkage of the continental lithosphere, occasioned by progressive cooling. Kenneth Deffeyes suggests the inevitability of this, after a continent has lost its most potassic and radioactive portion in being trimmed down to sea level.

Yet another phenomenon that offers a possible explanation is that of oscillating sea level. The stratigraphic record shows that sea level has oscillated markedly through geological time, relative to the continental surfaces, leading to an alternation between widely submerged (thalassocratic) and widely emerged (epeirocratic) states (Sonder, 1956; Belousov, 1962; Wise, 1972). Rona (1972) Pitman and Hays (per. comm.) and others have suggested that these oscillations may in large part derive from changes in the volume of the ocean basins, due to variations in the rate of lithospheric accretion and corresponding changes in the size of the oceanic rift bulge. At times of rapid accretion, the rift bulge is large and displaces more water than it does during times of slow growth. The rift bulge is presumably due to heat transferred to the lithosphere from the interior of the earth; its growth must correspond in some measure to shrinkage of the interior, which is presumably distributed to the lithospere as a whole, so that the growth of a rift bulge in parts of the oceanic lithosphere must be compensated by subsidence of continents and of other oceanic areas. This subsidence and the displacement of water from the oceans work together, flooding the continental platforms to produce the thalassocratic condition. Conversely, the stagnation of crustal accretion under the seas must lead to shrinkage of rift bulges and an increase in oceanic volume, which will drain the water off the continents. We have seen above that sea level is the ultimate control on the thickness of continental crust underlying the stable platforms. We postulate now that secular fluctuations in sea level will lead to excessive erosional thinning of the continental lithosphere during periods of unusual continental emergence, and to a restoration of this lithosphere by surficial sedimentation during periods of extensive flooding. Development of sedimentary blankets will be aided by the loading effects of water and sediment and leads to the development of a sedimentary cover over much of the continental platforms—a sedimentary cover which is broken by many interruptions (Sloss, 1964; Ham and Wilson, 1967; Wise, 1972).

SUBSIDENCE PATTERNS

We now turn from theoretical considerations to a pragmatic examination of subsidence patterns. Subsidence data for a variety of basins are here provided in the form of Bubnoff diagrams (Figs. 8, 9): plots of subsidence versus time, as pioneered by S. von Bubnoff (1954), and used by Belousov (1962), Fischer (1966), Garrison and Fischer (1969), and others.

One of these curves—that for new ocean floor in Fig. 8—represents the subsidence rate inferred for the flanks of midoceanic ridges, by plotting mean water depth against ages inferred from the associated magnetic anomalies. The other curves are all based on stratigraphic data: in most of them, the filling of a given basin consists of sediments deposited near sea level; thus, water depths may be ignored and the thickness of sediments accumulated at some given place, over a given time interval, is taken as a direct measure of subsidence. In some cases, however, the character of the sediments suggests the temporary development of appreciable water depths, and in these an inferred water depth has been shown. Among other sources of error are the following: (1) the curves are based on present (compacted) thicknesses. Since compaction followed deposition with some delay, the early subsidence rates are probably underestimated and the late ones overestimated. (2) The curves are drawn as straight-line segments between control points. These control points are distinctive stratigraphic boundaries such as the top of the Cambrian or the base of the Cretaceous, dated by a standard geological time scale (Harland, Smith, and Wilcock, 1964), but the age of some of these boundaries is not accurately known. Furthermore, such straightline segments can at best represent mean subsidence rates over the time intervals used, and may depart considerably from the actual subsidence patterns. (3) Our datum plane—sea level—is also subject to minor and greater oscillations. Minor variations are therefore probably not significant at the present level of knowledge.

An attempt was made to represent a wide variety of basins: a decaying rift bulge on new ocean floor; a rift valley fill (Klemme's Type 3); an interior platform; interior basins (autogeosynclines or synclises, Klemme's Type 1); Atlantic or rift margin type basins (Klemme's Type

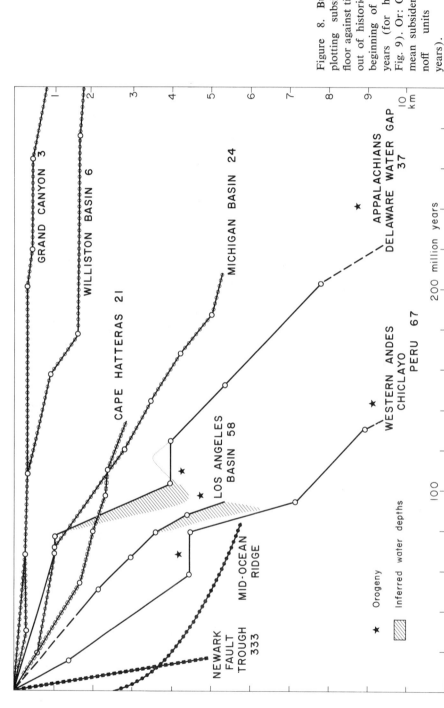

Figure 8. Bubnoff diagrams, plotting subsidence of basin floor against time. Curves taken out of historic content, setting beginning of subsidence at 0 years (for historic plot, see Fig. 9). Or: Orogeny; Figures: mean subsidence rate, in Bubnoff units (meters/million years).

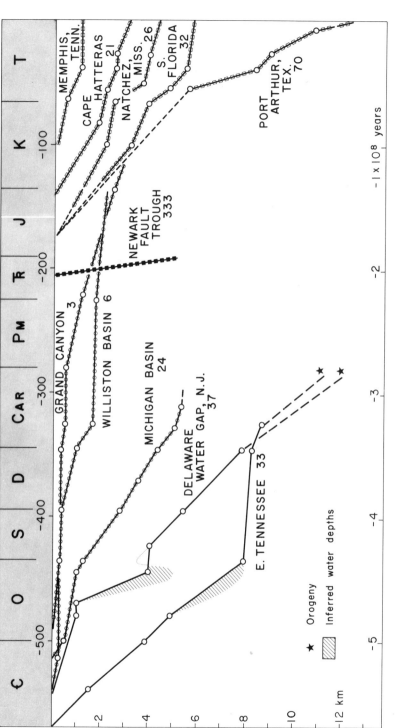

Figure 9. Bubnoff diagrams, plotting subsidence of basin floor against time. Curves plotted on historic base (for a different comparison, see Fig. 8). Or: Orogeny. Figure indicates mean subsidence rate, in Bubnoff units (meters/million years).

5); miogeosynclinal basins; a geosynclinal fill from an andesite belt; and a representative of orogenic successor basins (Klemme's Type 6). The two figures contain some of the same curves, and some limited to one or the other because of problems of overlap. Figure 9 has an historical base, whereas in Figure 8 the curves are taken out of their historical context and plotted with a common point of origin, the known or inferred beginning of subsidence.

These figures show a wide spectrum of behavior. At one extreme are the great interior platforms, showing mean subsidence rates of a few Bubnoffs: the Grand Canyon region (Eardley, 1951) persisted in this state for 260 million years, the Williston basin area (Perry, 1953) for 100, before becoming involved in more rapid sinking. This very slow and sporadic sinking I would attribute to the effect of sea level oscillation, discussed above, and is not really part of basin development, though such deposits come to be parts of basin fills.

In contrast to this, central parts of the Michigan basin (Dorr and Eschman, 1970), the type example of Kay's autogeosynclines, show a remarkably steady behavior through 200 million years of history, with a mean subsidence rate of 24 Bubnoffs. The center of the Williston basin showed this type of behavior during Devonian and early Carboniferous times, before settling down to another long period of stability. That was followed by a renewed period of marked subsidence, in the Cretaceous, not shown in our curves, and possibly related to the Laramide orogeny.

What caused the subsidence in these basins? In part it is surely a loading effect (Fig. 7): if we assume a mean density of the Michigan basin fill as 2.3, and an upper mantle density of 3.3, then the loading factor can account for more than two thirds of the basin depth—but not for the remaining 1.25 km, nor for the brisk pattern with which subsidence appears to have begun. Other factors started subsidence, and continued to contribute while loading assumed dominance. It is conceivable that one of these factors was the erosion and cooling of a transient heat bulge in the shield area over whose truncated roots the Michigan basin was formed. Alternatively or additionally other processes—an intrusion of heavy matter, a phase change at depth, or a change in the asthenosphere—may have been responsible.

While a truncated thermal bulge may have started the Michigan basin, it could not have set in motion the Devonian subsidence of the Williston basin, for this occurred long after subsidence and sedimentation had begun there at a very slow pace. Here we must reckon with one of the other possibilities. Whatever the cause, the effect was evidently smaller than that in the Michigan basin, and did not allow for either the subsidence rates or the continuation of load-induced subsidence that appeared to be recorded there, until some new factor once again initiated subsidence in Cretaceous time (not shown on our graph).

Turning from these long-dead basins to subsidence patterns on non-orogenic margins of North America, we have assembled a spectrum of curves—partly from the Gulf Coast, and partly from the Atlantic (Murray, 1961). In the updip areas, such as Memphis, Tennessee, or the New Jersey Cretaceous outcrop belt, we find low subsidence rates, initiated in mid-Cretaceous time and ending in the mid-Tertiary. Farther down-dip, as in the sections at Natchez, Mississippi, southern Florida, and Cape Hatteras, we find intermediate subsidence rates and a pattern which commenced briskly at Cape Hatteras; such a brisk beginning has been inferred for the other areas, where wells have not penetrated to basement. These sections not only subsided more rapidly, they started earlier, and they continued almost to the present; their present stability may be merely a byproduct of our monentary epeirocratic state. Finally, the sequence at Port Arthur, Texas, shows very much higher subsidence rates, continuing up to the present; its earlier history is totally inferential. Similarly, the thick sedimentary sequences discovered by seismic means off the Atlantic coast show the presence there of rapidly sedimented and even more rapidly subsided (drowned) basins.

As will be brought out in more detail by Kinsman, we interpret this marginal basin belt to be largely the result of a truncated thermal bulge associated with rifting, in Jurassic time. The updip limits of Cretaceous and Tertiary sediments all along the Gulf and Atlantic coasts were deposited on essentially unmodified continental crust; the intermediate section (Natchez, south Florida, Cape Hatteras) represents the subsidence of the truncated rift bulge, plus the added loading factor and perhaps other secondary effects. The thickest sequences, perhaps reaching land at Port Arthur but altogether off shore in the Atlantic, represent the maximum sedimentation, in the rift itself, deposited on attenuated continental lithosphere and on juvenile oceanic crust. The interpretation of the northern edge of the Gulf of Mexico as a rift margin, in direct parallel with the Atlantic coast may be unconventional, but Kinsman and I believe that it is supported by similarities in stratigraphy and structure.

The Triassic Newark series of New York, New Jersey, and Pennsylvania is deposited in strips of rift-attenuated continental crust, which most likely represent abortive predecessors of the Atlantic rifting. The Newark series is generally estimated to have a thickness of about 5 km, and its age, as interpreted from vertebrate faunas, is Late Triassic. It seems unlikely that its time span exceeded 15 million years, in which case the minimal subsidence rate is 330 Bubnoffs—the most rapid rate of any of the basins plotted here.

Turning now to basins in mobile belts, involved in orogeny, we examine first the miogeosynclinal curves for the Ridge and Valley Province. Subsidence in the Delaware Water Gap area of New Jersey (Epstein, 1969; Drake, 1969) proceeded very slowly up to mid-Ordovician times, in a manner reminiscent of the interior platforms or the inner margin of the continental margin basins, while subsidence commenced briskly in eastern Tennessee (King, 1959), even faster than the Mesozoic subsidence at Cape Hatteras. Then, in mid-Ordovician time, subsidence rates became very high in both areas, leading to basin starvation and the development of turbidites, the Martinsburg or Blount flysch facies. This sudden increase in subsidence rates coincides with the first influx of sediments (graywacke) from the east, marking a fundamental geological change coinciding with the Taconic orogeny. Later stages of this orogeny resulted in intense deformation and moderate uplift and beveling in the New Jersey area, whereas in Tennessee it was merely expressed in a cessation of subsidence.

The significant point to be made is that neither of these curves suggests, in the first 100 million years of their history, a mobile belt in which isostasy was violated: both look like continental rift margin basins. The slow subsidence at the Delaware Water Gap suggests the continental edge of the initial rift bulge, outside the area of denudation; more rapid subsidence in eastern Tennessee suggests a site on the rift bulge proper. This is in harmony with the picture developed by Bird and Dewey (1970) for the Northern Appalachians. Indeed, one wonders if the Catoctin spilites of the Blue Ridge may not represent basalts of a related rift valley.

The spurt of orogenic (Taconic) sinking may reflect the development of orogenic inhomogeneities in the asthenosphere. In the New Jersey area, a new episode of subsidence—slightly more rapid than that going on at this time in the Michigan basin—commenced in the mid-Silurian, and presumably continued into the Carboniferous. Another episode of massive subsidence may have occurred in the development of Carboniferous molasse basins, known from other parts of the Appalachian region, but here the record has been lost by erosion. However this may be, the area was then folded once again in the late Carboniferous-Permian Appalachian orogeny. Like the Cambro-Ordovician, this Silurian to early Carboniferous subsidence pattern shows no signs of belonging to a mobile belt: instead, it is part of a broad basin, the Allegheny synclinorium (Kay, 1942), which partly overlapped onto the bevelled Taconic fold belt, and was in turn partly incorporated into the Appalachian fold belt. One wonders whether this may not represent a lithospheric sag over a patch of asthenosphere modified by the Taconic orogeny.

Eastern Tennessee behaved like a continental platform through Silurian and Devonian times, and was then presumably involved in more intense Carboniferous subsidence, prior to Appalachian folding.

The Paleozoic Ardmore-Anadarko basin complex of the Mid-Continent region, not depicted in our illustrations, presents a particular challenge. Once considered as "the deepest and geologically the most persistent basin of the craton" (Ham, 1967), it is probably not a cratonic basin at all: The presence of a great mafic layered complex (Raggedy Mountain Gabbro) of Cambrian age, and of a spilitic basalt sequence (Ham, Denison, and Merritt, 1964; Ham, 1967) strongly suggests that this area was rifted in early to middle Cambrian time, developing possibly beyond the crustal attenuation stage to the emplacement of juvenile oceanic lithosphere, but stopping well short of the dimensions developed by the Red Sea. The Cambrian granites and rhyolites may be secondary features derived from the heated continental crust. If this be correct, then we should expect to find remnants of a central strip of juvenile lithosphere; flanked by belts of attenuated continental crust; in turn flanked by belts of truncation, showing a progressive sedimentary onlap (see Kinsman, this volume). The stratigraphy of the region needs review in the light of this possibility.

For a geosynclinal volcanic basin, I have used a section measured at Chiclayo, Peru, by I. Tafur and myself (Fischer, 1956). Here a basement of beveled, low-rank Paleozoic metamorphics sank rapidly in Late Triassic (Norian) time and through much of the Jurassic, accumulating a thick cover of intermediate volcanics with interbedded sediments ranging from shoalwater to continental. Dating in the middle part of the section is poor, and the presence of late Jurassic folding is not established, but such a folding, in pre-Portlandian time, is widely recognized in the Peruvian Andes; the interruption of subsidence, in our curve, is an interpolation. Vigorous subsidence continued through the Cretaceous, of which the early part is continental, and the later part is marine, this area being on the edge between the sedimentary carbonate-dominated facies to the east and the volcanic turbidite facies of the west. In Senonian or Maestrichtian time, the area was subjected to gentle folding, intrusion, and zeolite-grade metamorphism, the axis of the great batholith lying just to the west of it. Deformation is mainly by large-scale block faulting, of undetermined timing. This area is part of the great andesite belt extending from Chile to Colombia, that has occupied roughly the same position from late Triassic or Jurassic time to the present. The tectonic setting would thus appear to have been near the frontal edge of an overriding plate, perhaps 180 km from a trench (which would then have been near where it is today).

What caused this subsidence of an area that had shortly before been a mountain welt? The decay of a heat bulge seems unlikely, inasmuch as the volcanic activity here suggests continued and probably increased heat flow. Is it a belt of tectonic attenuation, an incipient intra-arc basin? Or is this an example of the heterogeneities introduced into the asthenosphere by subduction—perhaps an area of sublithospheric magma accumulation, into which the lithosphere foundered? Or is this a case in which magma chambers within the lithosphere have collapsed, transferring their fill to the subsiding surface above them? Here we are reduced to guesses, lacking criteria for testing hypotheses.

The last of the mobile belt basins here plotted is the Los Angeles basin, a successor basin developed on Franciscan basement (Natland, 1957). Its Cretaceous history is here inferred. Through the Paleogene it subsided rapidly, passing repeatedly from marine to continental sedimentation. In Miocene time the facies turned progressively deeper, reaching its most starved conditions in Pliocene flysch facies with attainments of water depths, according to faunal studies, in excess of 1.5 km. The ultimate phase of the basin's history involved rapid filling, terminated by orogeny in the Pleistocene. Various basins of California show a similar development, in synchrony. We understand neither the general subsidence of the Paleogene nor the causes for the extreme Neogene subsidence in several distinct but neighboring basins, but this again may reflect the formation of asthenospheric inhomogeneities. Alternatively, the transition from a convergent leading plate margin to a shear margin may well have brought about a fault-slivering of the lithosphere that caused tectonic attenuation in some areas.

CONCLUSIONS

A look at basins, from the viewpoint of plate tectonics reveals various classes, with distinct origins and behavioral patterns. The basins of first order, the oceans as such, are largely explained by rifting: they are divisible into second order transverse basins by original differences in oceanic crust, and overprinted on these are the basins that develop on either side of a crust-generating, rifting ridge as a result of cooling and perhaps phase changes. Marginal oceanic (interarc) basins are here attributed to the foundering of oceanic lithosphere.

The great basins developed on continental margins, linked to rifting oceans, are interpreted as initiated by the erosional truncation of the initial, temporary, thermal rift bulge, followed by cooling. In all basins, loading by water and by sediment plays a large role.

Other kinds of basins on continents include rift valleys that never grew into oceans, and other belts of tectonic stretching, expressed in complex block fault patterns and subsidence.

Certain areas of the continental interiors sink more than surrounding ones, to form the great interior basins, autogeosynclines or synclises. The initiation of these basins is problematic. Some may owe their origin to the subsidence of a truncated thermal bulge, others to phase changes at depth or to mantle differentiation. Loading by water and sediment is surely an important factor in their subsequent deepening.

The most varied basins are basins found in mobile belts. A belt such as the Appalachian Valley and Ridge province began as a basin of the trailing continental margin type, passed into an orogenic state in the Taconic episode, entered into a platform-autogeosynclinal state, and once again into an orogenic state in the Appalachian episode. The orogenic states are characterized by localized but rapidly subsiding basins, which may be related either to anisostatic downdragging in the vicinity of subduction zones (trenches), or to the development of heterogeneities in the asthenosphere, causing the lithosphere to rise in some places and to sag in others.

Volcanic basins may also arise from the development of magma chambers in the lithosphere and the collapse of such chambers with concomitant accumulation of bedded volcanic rocks on the surface.

In summary, plate tectonics and other geological insights provide a variety of plausible mechanisms for the development of basins. These mechanisms trigger each other and interact, so that no basin is due to one cause alone: loading is an important factor in all. Plausible models can be constructed for some basins, such as the primary ocean basins or the basins of trailing continental margins (Kinsman, this volume). In the case of most basins, however, the initiating causes remain uncertain. The basin dynamics of the future must establish on the one hand the kinematics of the basin, as recorded in its fill, and must, on the other, interpret this behavior in terms of the complex of contributing causes.

As Klemme will show, basins form in a wide range of thermal settings. The actual thermal regime in any one basin will be modified by patterns of subsidence and loading, and probably also by patterns of water escape. The pressure regime will likewise be largely determined by rates of subsidence and loading. Together, these control the rates of compaction, rates and patterns of fluid movements, and the development of internal structures (diapirs etc.). Each basin has its own pressure-temperature regime, and different classes of basins share some attributes of these regimes. Thus, the geological setting of basins, their manner of origin, and their mode of growth must play a large role in the accumulation of fossil fuels, and of other resources concentrated in the course of diagenesis. In this sense, the study of basins has only just begun.

REFERENCES

Adams, J. E., H. N. Frenzel, M. L. Rhodes, and D. P. Johnson, 1951. Starved Pennsylvanian Midland Basin (Texas), *Bull. Amer. Assoc. Petrol. Geol.,* v. 35, pp. 2600–2607.

Alvarez, W., 1972. Uncoupled convection and subcrustal current ripples in the western Mediterranean. In *Studies in Earth and Space Science,* ed. R. Shagam, Geol. Soc. Amer. Memoir 132, pp. 119–132.

Ampferer, O., 1906. Über das Bewegungsbild von Faltengebirgen. *Abh. d. k. k. Reichsanstalt* (Wien), v. 56, pp. 539–622.

———— and W. Hammer, 1911. Geologischer Querschnitt durch die Ostalpen vom Allgäu bis zum Gardasee, *Jahrb. geol. Reichsanst. Wien,* v. 61, pp. 531–710.

Anderson, D. L., 1962. The plastic layer of the earth's mantle: *Continents Adrift,* Scientific American, pp. 28–35.

Beloussov, V. V., 1962. *Basic Problems in Geotectonics,* McGraw-Hill, 809 pp.

Bird, J. M., and J. F. Dewey, 1970. Lithosphere-continental margin tectonics and the evolution of the Appalachian orogen, *Bull. Geol. Soc. Amer.,* v. 81, pp. 1031–1060.

Bubnoff, S. von, 1954. *Grundprobleme der Geologie,* Berlin, Akademie Verlag; edited translation by W. T. Harry, *Fundamentals of Geology,* Oliver & Boyd, 287 pp.

Bucher, W. H., 1933. *The Deformation of the Earth's Crust,* Princeton University Press, 518 pp.; reprinted Hafner, New York, 1959.

Cloos, H., 1939. Hebung, Spaltung und Vulkanismus: *Geol. Rundschau,* v. 30, Zwischenheft 4A, pp. 405–527.

———, 1951. *Gespraech mit der Erde,* R. Piper & Co., Munich, 309 pp.

Dallmus, K. F., 1958. Mechanics of basin evolution and its relation to the habitat of oil in the basin. In *Habitat of Oil,* ed. L. G. Weeks, Amer. Assoc. Petrol. Geol., pp. 883–931.

Daly, R. A., 1940. *Strength and Structure of the Earth,* Prentice-Hall, 434 pp.

Dorr, J. A., and F. F. Eschman, 1970. *Geology of Michigan,* University Michigan Press, 476 pp.

Drake, A. A., Jr., 1969. Precambrian and Lower Paleozoic Geology of the Delaware Valley, New Jersey–Pennsylvania. In *Geology of Selected Areas in New Jersey and Eastern Pennsylvania,* ed. S. Subitzky, Rutgers University Press, pp. 51–131.

Eardley, A. J., 1951. *Structural Geology of North America,* New York, Harper.

Epstein, J. B., and A. G. Epstein, 1969. Geology of the Valley and Ridge Province between Delaware Water Gap and Lehigh Gap, Pennsylvania. In *Geology of Selected Areas in New Jersey and Eastern Pennsylvania,* ed. S. Subitzky, Rutgers University Press, pp. 132–213.

Fischer, A. G., 1956. Desarollo geologico del Noroeste Peruano durante el Mesozoico: *Soc. Geol. del Peru,* v. 30, pp. 177–190.

———. 1964. The Lofer cyclothems of the Alpine Triassic, *Symposium on Cyclic Sedimentation,* ed. D. Merriam, Kansas Geol. Survey Bull. 169, pp. 107–149.

———. 1969. Geological time-distance rates: The Bubnoff unit. *Bull. Geol. Soc. Amer.,* v. 80, pp. 549–552.

Fischer, A. G., and B. C. Heezen, et al., 1971. *Initial Reports on the Deep Sea Drilling Project, v. 6, Honolulu, Hawaii, to Apra, Guam.* Govt. Printing Office, Washington, 1329 pp.

Garrison, R. E., and A. G. Fischer, 1969. Deep water limestones and radiolarites of the Alpine Jurassic. In *Depositional Environments in Carbonate Rocks,* ed. G. Friedman, Soc. Econ. Paleont. and Mineral., Spec. Publ. 14, pp. 543–572.

Gilluly, J., 1955. Geologic contrasts between continents and ocean basins. In *Crust of the Earth,* ed. A. Poldervaart, Geol. Soc. Amer. Special Paper 62, pp. 7–18.

———. 1964. Atlantic sediments, erosion rates and evolution of the continental shelf, *Geol. Soc. Amer. Bull.* v. 75, pp. 483–492.

Halbouty, M. T. et al., 1970. Factors affecting formation of giant oil and gas fields and basin classification. In *Geology of Giant Petroleum Fields*, ed. M. T. Halbouty, Amer. Assoc. Petrol. Geol. Mem. 14, pp. 528–555.

Ham, W. E., and J. L. Wilson, 1967. Paleozoic epeirogeny and orogeny on the central United States, *Amer. Jour. Sci.*, v. 265, pp. 332–407.

Ham, W. E., Denison, R. C., and Merritt, C. A., 1964. Basement rocks and structural evolution of southern Oklahoma, *Okla. Geol. Surv. Bull.* 95, 302 pp.

Harland, W. B., A. G. Smith, and B. Wilcock, eds., 1964. *The Phanerozoic time-scale*. Geol. Soc. London, v. 120s, 458 pp.

Haug, E., 1900, Les géosynclinaux et les aires continentales. Contribution a l'etude des transgressions et les régressions marines, *Soc. Géol. France Bull.*, 3rd ser., v. 28, pp. 617–711.

———, 1907. *Traité de Géologie*, Paris, Armand Colin, v. 1, 538 pp.

Heezen, B. C., 1960. The rift in the ocean floor, *Scientific American*, v. 203, p. 98–110.

Hess, H. H., 1955. Serpentines, orogeny, and epeirogeny. In *Crust of the Earth*, ed. A. Poldervaart, Geol. Soc. Amer. Spec. Paper 62, pp. 391–408.

———, 1962. History of the Ocean Basins. In *Petrologic Studies*, ed. Engel, James, and Leonard, Geol. Soc. Amer., pp. 599–620.

Hollister, C. D., Ewing, J. I., et al., 1972, *Initial Reports of the Deep Sea Drilling Project*, v. XI. Washington (U.S. Gov. Print. Off.), 1077 pp.

Holmes, A., 1965. *Principles of Physical Geology*. New York, Ronald Press, 1288 pp.

Hsü, D. J., 1965. Isostasy, crustal thinning, mantle changes and the disappearance of ancient land masses, *Am. Jour. Sci.*, v. 263, pp. 97–109.

Isacks, J. B., J. Oliver, and L. R. Sykes, 1968. Seismology and the new global tectonics, *J. Geophys. Res.*, v. 73, pp. 239–254.

Karig, D. E., 1971. Origin and development of marginal basins in the Western Pacific, *J. Geophys. Res.*, v. 76, pp. 2542–2561.

———, 1972. Remnant Arcs, *Bull. Geol. Soc. Amer.*, v. 83, pp. 1057–1068.

Kay, M., 1942. Development of the Allegheny synclinorium and adjoining regions, *Bull. Geol. Soc. Amer.*, v. 53, pp. 585–646.

———, 1951. *North America Geosynclines*, Geol. Soc. Amer. Mem. 48.

King, P. B., 1959. *The Evolution of North America*, Princeton University Press, 190 pp.

Kossinna, E., 1921. Die Tiefen des Weltmeeres. *Veröff. Meereskunde Univ. Berlin*, N.F., Reihe A, H. 9, p. 33.

MacDonald, W. D., 1972. Continental Crust, Crustal Evolution, and the Caribbean. In *Studies in Earth and Space Sciences*, ed. R. Shagam, Geol. Soc. Amer. Mem. 132, pp. 351–362.

Menard, H. W., 1969. The deep ocean floor, *The Ocean*, Scientific American, pp. 51–63.

Moberly, R., 1972. Origin of lithosphere behind island arcs, with reference to the Western Pacific. In *Studies in Earth and Space Sciences,* ed. R. Shagam, Geol. Soc. Amer. Mem. 132, pp. 35–57.

Morgan, W. J., 1965. Gravity anomalies and convection currents, *J. Geophys. Res.,* v. 70, pp. 6189–6204.

———, 1968. Rises, trenches, great faults and crustal blocks, *J. Geophys. Res.,* v. 73, pp. 1959–1982.

———, 1971. Convection plumes in the lower mantle, *Nature,* v. 230, pp. 42–43.

———, 1972. Deep mantle plumes and plate motions, *Bull. Amer. Assoc. Petrol. Geol.,* v. 56, pp. 203–213.

Murray, G., 1961. *Geology of the Atlantic and Gulf Coastal Provinces of North America,* New York, Harper, 692 pp.

Natland, M. L., 1957. Paleoecology of West Coast Tertiary sediments. In *Treatise on Marine Ecology and Paleoecology,* ed. H. Ladd, Geol. Soc. Amer. Mem. 67, v. 2, pp. 543–572.

Oxburgh, E. R., 1971. Plate tectonics. In *Understanding the Earth,* ed. I. G. Gass, P. J. Smith, and R. C. L. Wilson, Cambridge, Mass., M.I.T. Press, pp. 263–285.

Perry, E. S., 1953. Oil and Gas in Montana, *Mont. Bur. Mines and Geol.,* Mem. 35, 54 pp.

Rona, P. A., 1972. Relation between rates of sediment accumulation on continental shelves and sea floor spreading in the central North Atlantic, *Geol. Soc. Amer.* Abstr. v. 4, p. 644.

Schneider, E. D., 1972. Sedimentary evolution of rifted continental margins. In *Studies in Earth and Space Sciences,* ed. R. Shagam, Geol. Soc. Amer. Mem. 132, pp. 109–118.

Sclater, J. G., and J. Francheteau, 1970. The implication of terrestrial heat flow observations on current tectonic and geochemical models of the crust and upper mantle of the earth: *Geophys. Jour. Royal Astronom. Soc.,* v. 20, pp. 509–542.

Sloss, L., 1964. Tectonic cycles of the North American cration. In *Symposium on Cyclic Sedimentation,* ed. D. Merriam, Kansas Geol. Surv. Bull. 169, pp. 449–460.

Sonder, R. A., 1956. *Mechanik der Erde,* Schweizerbart, 291 pp.

Stone, G. E., ed., 1967. *Field Trip Guidebook. The Structures and Igneous Rocks of the Wichita Mountains, Oklahoma.* First Ann. Meet. South-Central Section, Geol. Soc. Amer. (dist. Okla. Geol. Survey).

Travis, R. B., 1953. La Brea-Pariñas Oil field, Northwestern Peru, *Bull. Amer. Assoc. Petrol. Geol.,* v. 37, pp. 2093–2118.

Wise, D. U., 1972. Freeboard of continents through time. In *Studies in Earth and Space Sciences,* ed. R. Shagam, Geol. Soc. Amer. Mem. 132, pp. 87–100.

Wyllie, Peter J., 1971. *The dynamic earth,* N. Y., John Wiley & Sons, 416 pp.

From basins in general, we now pass to David Kinsman's discussion of the development of rift basins, and of the basins flanking rifts on their continental margins (the "Atlantic type" basins of Curray). These basins, containing the greatest unexplored volume of prospective sediment, seem to find a particularly clear explanation in plate tectonics and the temporary heat bulges associated with rift zones. Their stratigraphy lends itself to prediction.

Rift Valley Basins and Sedimentary History of Trailing Continental Margins

DAVID J. J. KINSMAN[1]

ABSTRACT

A model of continental lithospheric rifting is proposed in which initial uplift and post-rifting subsidence are directly related to subcrustal lithospheric temperature and density distributions; the model is analogous to that proposed for sea floor rifting and subsidence. During all stages, isostatic equilibrium is maintained. Early domal uplifts over deep mantle plumes are followed by interplume uplifts as a linear rupture zone is developed. Rift valleys are formed between these divergently rifting continental margins and the elevated areas are erosionally thinned. The eroded, elevated areas, several hundred kilometers in width, later subside below sea level to form the continental terrace. Along transcurrently rifted continental margins, no uplift accompanies rifting and no erosional thinning occurs. A zone of major intracrustal attenuation, 30–40 km wide along transcurrent margins and 60–80 km wide along divergently rifted margins marks the outer edge of

[1] Department of Geological and Geophysical Sciences, Princeton University.

the continent. This zone comprises the 3–4 km high continental slope, which early in its history exists partly above sea level; as the underlying lithosphere cools, the continental slope slowly subsides below sea level. The distribution of crustal thicknesses across these two continental margin zones is reflected later in the thickness and extent of the sediments which are deposited.

A variety of conclusions is drawn from the model. It is suggested that rift valley basins should occur along some continental margins, in the zone of attenuation underlying the continental slope. The first marine sediments deposited on the continental terrace will be 30–60 MY older than the first marine sediments laid down on the adjacent juvenile ocean floor. Evaporites frequently occur in juvenile rift oceans, lying on newly formed oceanic crust and on attenuated continental margin crust. Continental fragments, continents rifting along transcurrent segments, or volcanic islands over plume sites form effective evaporite basin barriers during early ocean evolution. The sediment loading capacity of oceanic crust increases with time but rarely exceeds 16–18 km; this is also the typical sediment thickness found for reconstructed ancient eugeosynclinal belts. Early evaporites will lie at the base of the thickest pile of sediments.

INTRODUCTION

Trailing, rifted or Atlantic type continental margins are very extensive today, bordering nearly the entire north and south Atlantic Ocean, much of the Indian Ocean, the Red Sea, and possibly much of the Arctic Ocean. These margins range in age from late Triassic/early Jurassic to late Tertiary/Quaternary. Along trailing continental margins a variety of sedimentary accumulations occurs; some of these are relatively thin, whereas others are many kilometers in thickness; some are laterally restricted, whereas others are laterally extensive. The more extensive, thicker deposits might be described as trailing margin sedimentary basins. This paper will explore some aspects of the sedimentary history of trailing or rifted continental margins within a sea-floor spreading/plate tectonics framework of earth structure and history. As the earliest stage of continental break-up is accompanied, at least sometimes, by rift valley development, sedimentary accumulations in rift valleys must be considered, in addition to the sediments that accumulate upon the margin of the rifted continent and on the adjacent ocean floor.

To accomplish these aims, it is necessary to discuss some elements of crustal and lithospheric structure; possible causes of elevation and subsidence, thinning, rupture, and response to loading and unloading. I then present an admittedly speculative model of continental rifting, but one that seems internally consistent and has, if no other merit, the possibility of being tested; the model has survived initial tests, some of which are described. The model permits recognition of two different types of rifted continental margin, which are distinguished by contrasting tectonic and sedimentary histories.

The outer 100 km of Earth is considered to comprise a fairly rigid shell, the lithosphere overlying a more plastic asthenosphere. The upper, more brittle part of the lithosphere is loosely termed *crust,* of which there are two major types, the essentially granitic continental crust and the essentially basaltic oceanic crust. The lithosphere is broken into a small number of fairly well-defined, large *plates* and a rather greater, but as yet poorly defined number of small plates. A single plate, or any discrete length of its periphery, may comprise continental lithosphere, oceanic lithosphere, or both types together. Plates move relative to one another and three broad types of plate boundary are recognized, dependent on the relative motions of the plates at these boundaries. Trailing or rifted plate boundaries are developed where the motion between two plates is divergent; differentiated mantle material wells into the axial rift zone between the plates, forming juvenile oceanic crust. Leading plate boundaries are developed where two plates converge (or founder; see Fischer, this volume), one plate plunging into the asthenosphere. A third type of plate boundary occurs where the motion between two plates is essentially a shearing or transcurrent motion. (The term *margin* is used here to describe the rifted edge of a continent).

Due to its lower density, continental crust tends to ride high, is not appreciably consumed where it forms leading plate boundaries, and makes up most of the crustal areas above sea level. In contrast, the denser oceanic crust floors much of the ocean, beneath a veneer of sediment, is generated at axial rifts, and is ingested into the asthenosphere at subduction zones, at horizontal rates of a few cm/yr. The mobility and ephemeral nature of the oceanic crust is reflected by the general youthfulness of the ocean floor (<200 MY), in contrast to the preservation of much older crust within the continental areas.

The segmental character of the lithospheric shell and the motions occurring between segments leads to a concentration and intensification of geologic events along plate boundaries—events such as igneous activity, tectonic deformation, and sedimentation. These phenomena are particularly concentrated at leading plate boundaries, but, nevertheless, a significant tectonic and sedimentational history can be demonstrated for trailing plate boundaries. (They are not places "where nothing happens" as was suggested, tongue-in-cheek, by Sir Edward Bullard during presentation of these papers!) The interiors of plates, particularly if of continental crust, have a rather more quiescent, dignified, geological existence.

This conceptual and terminological framework is essentially that used in the preceding paper by Fischer and more widely in the recent geological literature (Bird and Isacks, 1972). An additional concept, basic to some elements of this paper, is that of deep mantle plumes (Morgan, 1971), which are here considered to be responsible for the inception of continental rupture. It is within this framework that I shall discuss the origin and development of sedimentary basins. Perhaps "sites of accumulation of significant thicknesses of sediments" would be a better way in which to view the succeeding discussion; we call these sedimentary basins, but only in retrospect. During their accumulative phases, it would be difficult to conceive of many of them as basins, if we use the latter term in its more usual, popular sense.

If sediment accumulations are viewed in their relationship to lithospheric plates, then a division into plate boundary and plate interior may be usefully made. Two types of relatively quiescent plate interiors may be recognized, related to fundamental crustal differences: (a) oceanic crust, where a veneer of pelagic, volcanic, and fine-grained terrigenous sediments slowly accumulates, to be ultimately added to the leading edge of a plate, within a time span of <200 MY, and (b) continental crust, where cratonic basins accumulate a variety of terrigenous, carbonate, or evaporite sediments over extended periods of time (often >200–300 MY) and live to a ripe old age (>500 MY in many instances). Sediment accumulations of plate boundaries are more complex: (a) leading edge accumulations include tectonically emplaced oceanic sediments derived from the subducted plate together with the largely terrigenous sediments shed from the adjacent uplifted continental areas; (b) continental rift valley and continental trailing margin sediment accumulations, which form the main subject of this paper. As rifting of continental crust and formation of juvenile oceanic crust proceed, the rifted continental margin locale is incorporated into the plate interior, coming to lie progressively further within it. In other words the rifted continental margin is a plate boundary only in the initial stages. However, the later as well as the early tectonic and sedimentational history of these rifted continental margins distinguishes them from other plate interior sites of sediment accumulation. At continental margins sedimentation occurs across the junction between the two fundamental types of crust, continental and oceanic, which have very different sediment loading capacities. Oceanic crust adjacent to trailing continental margins commonly has very thick piles of sediment accumulate upon it, so thick in some instances that the loading capacity is fully taken up, and the top of the sediment pile lies at sea level.

The fundamental hypothesis presented here is that *the uplift, rupture, and later subsidence of divergently rifted continental margins are controlled in the first degree by the thermal history of the underlying lithosphere and asthenosphere during these events* (in an analogous way to that hypothesized for rifted oceanic crust by Sclater and Francheteau, 1970, and McKenzie, 1967). This then provides the major constraints on associated sedimentation, controlling such variables as maximum thickness and horizontal extent of sediment accumulations, stratigraphy of sediments across the continental margin, size of sediment source areas, and likelihood of evaporite formation.

Some elements of this proposal have been entertained earlier by several authors. For example, Hsü (1958, 1965) accepted the idea that thinned continental crust could later subside below sea level and that sediments could then accumulate on this sialic floor. He rejected the concept of crustal thinning by subcrustal processes and suggested that thinning could be achieved entirely by a variety of supracrustal processes such as erosion, of those continental regions that had been isostatically uplifted in response to density decreases in underlying mantle materials. Later increase in mantle density would bring about crustal subsidence. Crustal thinning and geosynclinal sedimentation were apparently considered by Hsü to be essentially intracontinental phenomena. The disappearance of older land masses that had served as sediment source areas at one time, such as Atlantis, which had supplied sediment from the east to the Appalachian geosyncline, was explained by subsidence beneath sea level following extensive supracrustal thinning (vertical motions of the crust being caused by mantle density changes). A spatial, temporal, and phenomenological framework, within which the suggested changes would occur in subcrustal mantle density, was not entertained by Hsü. One should note that his paper largely predated the concepts of plate tectonics and sea floor spreading.

Hess (1960) considered that the eugeosynclinal sediments of deformed alpine mountain systems accumulated in areas of thin continental crust along continental margins. In a series of papers (1963, 1965, 1966), Dietz discussed sediment accumulations along continental margins. In contrast to Hsü, he suggested that eugeosynclinal or continental rise sediments were deposited on oceanic crust and were ensimatic rather than ensialic. The coastal plain wedge or miogeosynclinal sediments were suggested to be ensialic. The transition from continent to ocean was not discussed by Dietz, and he apparently did not consider the possibility of the thinning of the edge of the continent and the relationship of thinning to subsidence. The continent seemingly was considered to terminate abruptly. No vertical isostatic response to subcrustal density variations was discussed by Dietz; he thought that "given initially the ocean floor, a continental platform, and topographic contrast of circa 4–5 km, then a continental rise sedimentary prism can accumulate to a thickness of circa 12 km before it overlaps the shelf break. Under this load I suppose that the continental terrace is sympathetically downflexed, since isostatic compensation is regional and not local" (Dietz, 1965). Dietz also considered the sialic and simatic areas to be strongly coupled throughout the major miogeosynclinal-eugeosynclinal sediment accumulation stage; after a thick pile of sediments had accumulated, uncoupling occurred, and the sediment prism was deformed.

Rona (1970) found that sediment accumulation data, from opposing north Atlantic continental margins, point to very similar histories of subsidence. He interpreted this to mean that the continental margins have behaved as though they were vertically and horizontally coupled to the ocean basin. The subsidence of the continental margin was found to vary coevally with sea floor spreading rates, determined for the floor of the intervening ocean, and Rona suggested a genetic relationship between rates of marginal vertical movement and rates of horizontal spreading of the sea floor. In a subsequent paper (Rona, 1972), the periods of faster sediment accumulation were suggested to be related to eustatic sea level changes brought about mainly by an increase in volume of the world ocean ridge system, which would be related to periods of more rapid sea floor spreading.

An important stratigraphic corollary follows from my earlier proposal, which relates sedimentation history to the thermal history of subcrustal lithosphere beneath a rifted continental margin. The post-rifting history of the continental margin will be partly reflected by an extensive time lapse between the earliest marine sediments accumulated on the oldest oceanic crust and the earliest marine sediments accumulated on the terrace or upper surface of the continental margin. The extent of this time lapse will be dependent on the amount of crustal thinning that has taken place, but can be shown to be of the order of 30 to 60 MY. Since these ideas were formulated, a paper by Sleep (1971) has been brought to my attention, in which a somewhat similar proposition has been made, to interpret some aspects of the sedimentary history of the Atlantic margins of North America.

Although many generalizations are used in the following pages, the basic model seems to fit a surprisingly large amount of fragmentary data on continental margin sediment ages and distributions. What seem to be reasonable speculations are entertained in several places, but I have tried to justify them as much as possible. I do not pretend that our understanding of the sedimentary history of continental margins is complete, or that our view of this problem is clear and transparent, but at least translucency may have replaced the former opacity.

The sedimentary history of rifted continental margins aroused my curiosity during a graduate sedimentology seminar in the fall of 1971, and I acknowledge the stimulation and discussion of all who participated; I would particularly thank Alfred Fischer and Christine Powell for their continuing interest, patience, and guidance.

TOPOGRAPHIC ELEVATION, CRUSTAL STRUCTURE
AND ISOSTASY

Surveys of Earth surface elevation by Kossinna (1921) and of ocean floor elevations alone by Menard and Smith (1966) have demonstrated that two major levels may be identified: (a) one relatively close to sea level, which comprises the greater part of the continental crustal areas (>75% of continental crustal areas lies between +1.5 and −0.5 km, relative to sea level); and (b) one between 3 and 6 km below sea level, which comprises the greater part of the oceanic crustal areas (>80%). These elevation data and associated crustal structures and their interpretation have been discussed by Holmes (1965), Hsü (1958, 1965), and many others. A modified version of the two most typical lithospheric columns is shown in Figure 1.

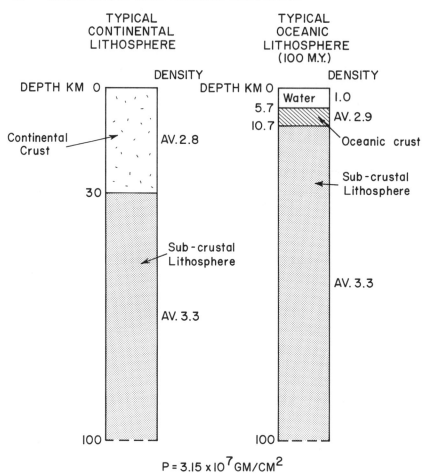

P = 3.15 x 10^7 GM/CM^2

Figure 1. Lithospheric columns beneath typical continental and oceanic crust. The oceanic column is for ocean floor about 100 MY in age. Crustal thickness and rock densities used throughout the discussion are also shown (modified, after Holmes, 1965).

A fundamental premise is that of isostatic compensation: at some depth all columns exert an equal pressure. I have selected 100 km as this surface of equal pressure, as this is thought to coincide fairly closely with the base of the greater part of the lithosphere, but the exact level chosen is immaterial for our purposes. If we use the rock density values of Figure 1, all columns exert a pressure at 100 km of 3.15×10^7 gm/cm^2.

The oceanic lithospheric column shown in Figure 1 is isostatically compensated when loaded with 5.7 km of water. However, juvenile ocean crust is formed at spreading axes at a variety of shallower depths, but most commonly between 2 and 3 km below sea level (Sclater and Francheteau, 1970). These authors, and many others, have demonstrated that ocean crust subsides monotonically with increasing age (Fig. 2). The subsidence is grossly independent of spreading rate, and subsidence curves can be fitted fairly well to theoretical curves of steady state cooling of the lithospheric plate. In fact, in the Indian Ocean, bathymetry can be used as a good guide to age of the ocean floor (McKenzie and Sclater, 1971). Thus, because of decreasing lithospheric temperature, and therefore increasing rock density, with passage of time the isostatically compensated depth of the upper surface of a column of oceanic lithosphere changes from an average close to -2.5 km at near zero age to -5.0 km at 50 MY, to -5.5 km at 80 MY, and to -6.0 km at 150–200 MY. The major subsidence is completed within the initial 40–60 MY (Fig. 3). However, if excess volcanism occurs and a much thicker than average oceanic crust is formed (cf. Iceland and the Wyville Thompson Ridge; Bott et al., 1971), it will start its life at a higher elevation than -2.5 km, and at any particular age will have an isostatically compensated elevation higher than that of oceanic crust of average thickness but of the same age. The time-related subsidence history of all oceanic crustal columns is similar, differing only in initial and final elevations. Oceanic crustal elevations are thus subject to two major controls (omitting, for the present, sediment loading), the first being the thickness of the low density (~ 2.9 gm/cm^3) oceanic crust itself, the second being time-dependent changes in deeper lithospheric density distributions.

The dominating continental crust level is thought to be underlain by an average crustal thickness of 30 km, of average density 2.8 gm/cm^3, buoyantly overlying lower lithospheric material of density 3.3 gm/cm^3 (Fig. 1). Density variations do occur through the crust, upper level sedi-

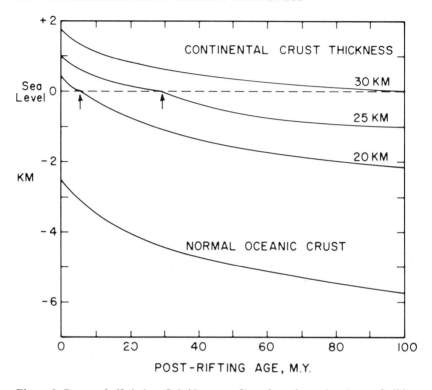

Figure 2. Decay of rift bulge: Subsidence profiles of continental and oceanic litho-spheric columns. The normal oceanic crust curve is modified from Sclater and Francheteau (1970); the continental crustal subsidence curves were calculated assuming subcrustal lithospheric densities at any time to be the same as those of suboceanic crustal lithosphere of the same age. The arrows mark the post-rifting time at which these continental columns reach sea level and begin to be loaded with sea water.

mentary rocks being in the range of 2.4–2.6 gm/cm³, upper metamorphic and igneous rocks averaging 2.7–2.8 gm/cm³, and lower crustal rocks perhaps averaging around 2.9 gm/cm³. In spite of obvious local variations in rock density, on a gross scale, 2.8 gm/cm³ seems to be fairly representative for average continental crust; unless otherwise stated, this average density will be used in subsequent discussions.

Continental elevations are explainable in terms of either variations in crustal thickness or in deeper lithospheric density distributions similar to those discussed for oceanic crustal columns. A third reason sometimes forwarded is bulk density increase of the continental crust, related for

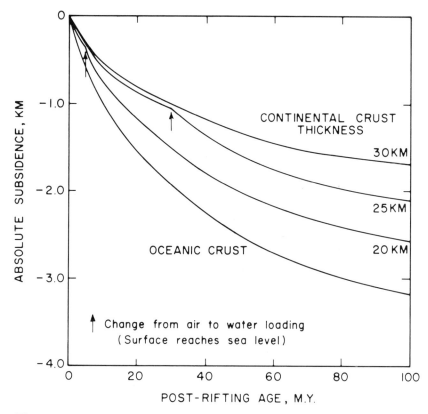

Figure 3. Decay of rift-bulge. Absolute subsidence-time curves for oceanic and continental lithospheric columns of 30, 25, and 20 km crustal thickness. The difference in subsidence between lithospheric columns at any age represents the amount of vertical uncoupling that would have occurred between adjacent litho-spheric columns at a continental margin. For example if an unthinned average continental lithospheric column is present at the continental margin, juxtaposed against an oceanic lithospheric column, then after 100 MY the differential sub-sidence between the upper rock surfaces totals 1.7 km (if no sediment loading has occurred).

example to intrusions of rock of density >2.8 gm/cm³. However the density of most basic intrusives is only ~2.9 gm/cm³. If the bulk of basic intrusives is emplaced within the deeper 10–20 km of the crust, the density contrast is rather small (2.9 vs. 2.8–2.9). The volumes required to achieve a significant mass change are very large and can rarely be demonstrated either to exist or to be likely. But if undifferentiated mantle materials of density 3.3 gm/cm³ were emplaced within the lower crust then the isostatic response could be relatively large for only fairly small volumes of intrusives. Emplacement of differentiated basic materials (~2.9 gm/cm³) into the upper levels of the crust where sedimentary rocks are commonly abundant and rock densities are in the range 2.6–2.8 gm/cm³ could generate marked loading effects, although piles of extrusives have a vesicularity and gross porosity that may well reduce their bulk density to values near those of other upper crustal rocks.

Crustal thinning, by whatever means, can readily account for the occurrence of sub-sea level continental crust. If a 30-km thick continental crust, isostatically compensated and lying initially close to sea level, is progressively thinned, the following water depths would be developed above it, using the model of Figure 1.

Continental Crust Thickness (km)	Overlying Water Depth (km)
30	0
25	1.1
20	2.2
15	3.2
10	4.3
5	5.4

Much of the ocean floor and the lithosphere beneath is younger than 80 MY. Heat loss over this 80-MY period causes the sea floor to change in elevation a total of 3 km; subsequent rates of heat loss and subsidence become ever smaller. As so little of the continental crust is underlain by lithosphere younger than 80 MY, continental elevation differences must presumably be related mainly to variations in crustal thickness. However, where continental crust is associated with lithospheric rifting at active rift valleys and at continental margins bordering very young oceans, then average thickness continental crust, riding buoyantly above a relatively low density subcrustal lithosphere, will lie appreciably above sea level. Elevations of 1–2 km might be predicted (Fig. 2) if the subcrustal lithospheric density distributions and associated thermal anomaly were similar to those under an oceanic rifting axis. With increasing age, if no crustal mass changes occurred (no thinning and no change in crustal bulk density), then the crust would subside and come to lie close to sea level once again after about 80–100 MY, as shown in Figure 2. Also shown in Figure 2 are subsidence curves for continental crust thinned to 25 km and to 20 km at the rupture stage and maintained at this thickness and constant density for 80–100 MY. In a young ocean, where juvenile oceanic crust is formed adjacent to a rifted continental margin it is proposed that *adjacent oceanic and continental crustal columns follow roughly similar subsidence-time curves,* as shown in Figure 2, at least for the initial 100 MY after rupture.

However, over this 100-MY period, an oceanic crustal column subsides a total of 3.2 km, from −2.5 to −5.7 km, whereas a continental crustal column (30 km thick crust) subsides only 1.7 km, from +1.7 to 0 km (Fig. 3). This difference is related to the differing densities of air and water and their loading effect in the above- and below-sea level situations. The difference in absolute amount of subsidence between the two isostatically compensated columns is 1.5 km, and this would require that the adjacent oceanic and continental crustal columns were not laterally coupled. Note, however, that the more extensive the thinning of the continental margin, the smaller becomes the difference in subsidence between adjacent oceanic and continental crustal columns. Lack of vertical coupling is even more emphasized when the sediment loading capacities of continental and oceanic crustal columns are contrasted across the continental margin; the net effect is to increase the subsidence contrast between the two columns to 10 or even 15 km. These conclusions are at variance with conclusions of Dietz (1965), Rona (1970), and Sleep (1971), who consider adjacent continental and oceanic lithosphere to be strongly coupled.

It should also be pointed out that, during post-rifting subsidence history, the continental and oceanic crustal columns we are discussing are *both* parts of the same lithospheric plate, not different plates as suggested by Sleep (1971).

There may be a significant difference in the later subsidence histories of oceanic and continental crustal columns. Oceanic crust continues to subside, after 100 MY, but at an ever decreasing rate; yet it is not at all clear that continental crust continues to undergo a similar continuing subsidence. It may be that heat loss has become so slow that sources of heat generation within the continental crust assume importance relative to simple loss of heat from the subcrustal lithosphere. In any event, the sedimentary evidence is difficult to interpret, for when variable short-period eustatic oscillations are superimposed on such slow subsidence, then the resulting picture becomes confused. It is possible that there is a long term net subsidence of stable cratons and that this may account for example for the widespread thin cover of older sedimentary rocks (>150 MY) over these cratonic platforms.

Two important aspects of any proposition concerning isostatic read-justment are (a) the rate at which compensation can occur relative to the rate at which disequilibrium processes are operating, and (b) the degree of isostatic disequilibrium that can be maintained before adjustment occurs. The first may be restated more directly in the following way: how rapidly does the crust respond to loading, for example by sedimentation or by water or ice addition, or unloading, for example by rock erosion or by water or ice removal? Glacial data from Lake Bonneville (Crittenden, 1963) demonstrate an isostatic response to un-loading of 100 m of water; over 90% of the readjustment was achieved in 10^4 years, the vertical crustal rebound rate being $\sim 10^4$ mm/10^3 years. Data for Scandinavia (Gutenberg, 1941) indicate an isostatic response time for 90% readjustment of 10^4–10^5 years and vertical rebound rates of 10^3–10^4 mm/10^3 years. Recent data from Greenland suggest 3×10^3 years for $\sim 90\%$ isostatic recovery and rebound rates of 10^4–10^5 mm/ 10^3 years (Ten Brink, 1972). These data are relevant for the present discussion because they demonstrate isostatic response times to be much shorter and associated rates of vertical motion to be much faster than nearly all rates of erosional unloading and sedimentational loading. *The conclusion, then, is that during both erosion and sedimentation epi-sodes rifted continental margins will be isostatically compensated.* The Lake Bonneville data also prove that the continental crust is capable of isostatic response to a load change as small as 10 bars (100 m of water or about 40 m of sediment). Isostatic equilibrium has been dem-onstrated over the Mississippi (Lawson, 1942) and Niger deltas (Hospers, 1965), where sediment thicknesses in places exceed 10 km and where adjacent continental and oceanic crust have responded isosta-tically to loading.

A further question might be asked concerning the way the crust re-sponds to loading and unloading. Is the response localized or do vertical movements extend well beyond the stressed area? This problem has been discussed recently by Crittenden (1963), Bloom (1967), Brotchie and

Silvester (1969), Walcott (1970, 1972), and Ten Brink (1972), and basically concerns the flexural rigidity of the crust. The amplitude and wavelength of flexure of a plate are functions of the flexural rigidity and also of the difference in densities of the material above and below the plate; the wavelength of flexure under water or sediments is longer than if the flexure occurred in air. Estimates of the lateral extent of vertical motion of continental lithosphere beyond the edge of a stressed area range from 50 to 200 km. Brotchie and Sylvester (1969) suggest that the flexural rigidity of oceanic lithosphere may be about one third that of continental lithosphere, although one might guess that this value changes markedly with age of the ocean floor, being related largely to changing thickness of the lithosphere with time.

The nature of the continent to ocean transition across the continental margin and the associated changes in flexural rigidity will drastically affect hypotheses of sediment accumulation across these margins. For example, if no change in flexural rigidity occurs across the continental margin, then the outer edge of the continental crust should be deeply downwarped by the extremely thick pile of sediments that can accumulate on oceanic lithosphere (see below), and the lateral extent of the downwarped area could be as great as 250 km (Walcott, 1972). If this relationship is true, then we should find a direct correlation between the thickness and the lateral extent of the miogeosynclinal sediment pile and the mass of sediments making up the continental rise or eugeosynclinal pile; it would be the loading effect of the latter sediments that would control the vertical and horizontal flexural downwarping of the miogeosynclinal region. A brief survey suggests that there is no direct correlation between either thickness and lateral extent of miogeocynclinical sediments and these parameters and mass of the eugeocynclinal or continental rise prism of sediments. However, this conclusion is considered to be no more than very tentative. The asymmetry of continental rise sediment prisms, steeper toward the continental margin, suggests either a change in flexural rigidity across the continental margin or a decoupling of oceanic and continental crust and subcrustal lithosphere.

The distribution of large fractures may be of great importance, permitting vertical adjustments to be limited to individual prisms and thus confining adjustment locally to the stressed area; evidence of this occurs in the development of down-to-basin growth faults. Continental rift valleys are typically defined by major faults and a system of subparallel faults sometimes extends beyond the confines of the rift valley itself.

The presence of major faults subparallel to rifted continental margins may lead to isostatic compensation's being largely localized to the areas loaded by sediment. But if faults are sparsely developed during the early rupture stage or are later annealed, flexure of the crust may extend well beyond the area of sediment loading. These two rather contrasting isostatic adjustment patterns may be important in the history of sedimentation on the continental margin itself. In this paper, however, no features less than 10 km in horizontal extent are considered.

CRUSTAL THINNING AND CRUSTAL RUPTURE

From the preceding section it follows that the evolution and distribution of crustal types and their thicknesses are signal controls on associated sedimentation. The temporal and spatial distribution of thinned continental crust will in part control the stratigraphies, facies, and thicknesses of sediments deposited along continental margins.

Crust may presumably be thinned by brittle fracture and may pull apart, perhaps plastically by necking or attenuation, or by removal of materials from upper or lower surfaces. The mechanics of lithospheric and crustal rupture have been much discussed, but little consensus seems forthcoming at this time. Even the marginal configurations of the ruptured continental crust are poorly known. To what extent crustal thinning precedes, accompanies, or post-dates rupture is likewise not known, and very different hypotheses have been advanced from various quarters.

Normal faulting of the upper, more brittle parts of the continental crust will bring about crustal thinning. If the dip of the fault planes decreases with depth, considerable thinning could be achieved in this way. Normal faulting presumably gives way at deeper levels to more plastic attenuation of the crust. Such intracrustal thinning cannot, however, be readily distinguished from subcrustal thinning. Subcrustal thinning has been extensively debated (see for example, Gilluly, 1964), but as yet there is little direct evidence to prove that the process operates. Sleep (1971) has remarked that, if subcrustal mantle currents are capable of melting off lower crustal material, they can do so only during the early updoming and rifting stages before the crust is underplated by cooling lithospheric mantle materials.

Rift valley basins are floored by blocks of continental crust, yet there is no evidence that they were ever elevated horst blocks; thus supracrustal erosional thinning cannot be responsible for their subsidence. Is it possible that the shape of the axial block controls the subsidence; if the block narrows downward, then this becomes a form of intracrustal thinning and subsidence must ensue. The axial blocks subside early in the updoming and rifting episode, and it would seem that intracrustal or subcrustal thinning processes must be called upon to account for their subsidence. Loading by basic igneous extrusives or intrusives is insufficient to account for the subsidence, as previously discussed. The recorded sediment accumulation rates for rift valley basins of 0.3 to more than 1 km/MY. (Van Houten, 1969; Zak and Bentor, 1972) require that crustal thinning proceeds at roughly similar rates during this early rifting stage.

Surficial processes of crustal thinning, or normal rock erosion, are obviously limited in their capacity to thin. Critical questions relate to the rates of erosion and the time available for erosion. If the history of continental updoming and post-rift subsidence presented earlier, and in Figure 2, is correct, then the time available for erosion, after onset of active spreading, is about 90–100 MY for average (30 km) continental crust. To this must be added the time that elapses between initiation of the sublithospheric thermal event (mantle plume inception; Morgan, 1971) and ultimate rupture of the crust. This time period is poorly known and may well be dependent on properties of the plume and overlying lithosphere and the larger scale stress relations between adjacent plates; in short, all those things that may permit or constrain lithospheric rupture. It would seem, on admittedly slender evidence (Baker et al., 1972), that this time period may be as small as 5 MY or as large as 20 MY. A third time element is related to the initial crustal thickness and to the total erosional thinning, as these in part determine the time when the upper surface of the continental crust subsides below sea level (Fig. 2), and thus can be no longer thinned by erosion. As an example, consider a 30-km crust erosionally thinned to 25 km, but isostatically adjusted for this thinning; instead of the upper surface reaching sea level after 90–100 MY, as would 30 km thick crust, it will reach sea level after 30 MY, a shortening of the potential erosion period by 60–70 MY. Thus the time available for erosional thinning can be as great as 120 MY, and is extremely sensitive to (a) the amount of crustal thinning achieved during the initial few tens of MY after updoming and rupture, and (b) the initial crustal thickness.

Literature on erosion has been accumulating rapidly, yet a concise summary acceptable to all protagonists is not easy to find. There is, however, a measure of agreement on the important role played by topography, climate, and rock type. The topographic factor is primarily that of relief, which is strongly related to gross elevation. Erosion rates have been stated to vary linearly (Ruxton and McDougall, 1967) or exponentially (Schumm, 1963) with relief. Climate has been shown by Langbein and Schumm (1958) to play an important role in controlling rates of erosion; rates are at a maximum when precipitation is in the range 25–30 cm/yr. At lower precipitation rates, run-off is apparently insufficient to transport debris away, and, at higher precipitation rates, the development of a cohesive vegetation cover over the land surface reduces particulate transport. The denudation rate varies by a factor of three between precipitation values of 25 and 50 cm/yr. The third factor, rock type, covers all those variables that describe the erodability of a rock; in general, sedimentary rocks are eroded more rapidly than igneous and metamorphic rocks.

Estimates of erosion rates by Schumm (1963) suggest 900 mm/10^3 yr under fairly high relief conditions, decreasing to values one tenth of this after relief has been largely destroyed. Judson and Ritter (1964) determined rates, ranging from 40 to 170 mm/10^3 yr, for large drainage basins in the United States, but averaging 60–70 mm/10^3 yr for the entire country. However, Meade (1969) and many others consider that erosion rates determined from studies of present day particulate and dissolved loads of rivers to be all badly affected by man's activities. How representative are such rates and how applicable they are to the geological past is presently uncertain.

Menard (1961) and Gilluly (1964) have made long-term erosion estimates for periods of 10^7–10^8 years by estimating the volumes of sediments that have accumulated over known time periods and by judging from paleogeographic reconstructions the extent of the sediment source areas. The method is far from precise but gives long-term rates of regional erosion of 20–60 mm/10^3 yr for the Appalachian and Mississippi regions and 210 mm/10^3 yr for the Himalayas.

Assume an updomed and newly rifted continental margin with a crustal thickness of 30 km. If exposed to an average erosion rate of 50 mm/10^3 yr, the surface of the crust would be reduced to sea level and its thickness to 27.5 km in a matter of 50 MY. If an average erosion rate of 170 mm/10^3 yr operated for 30 MY on such an updomed newly rifted continental margin, 5 km of erosional thinning would occur before the surface subsided below sea level (see Fig. 2). If the initial crustal thickness was greater than 30 km, a particular erosion rate would have to operate for a longer time before the surface subsided below sea level. The distribution of erosional thinning will vary across the updomed area. The belt nearest the rifted margin itself will have the highest elevation and relief and will be underlain longest by low-density subcrustal materials. It will therefore be subjected to the highest erosion rates, for the longest period of time, will undergo greatest erosional thinning and, ultimately, the greatest subsidence.

The time delay between rifting and subsequent marine flooding of the upper surface of the erosionally thinned, rifted continental margin (the future continental terrace) is dependent on (a) variations in the initial crustal thickness, (b) the erosional thinning that occurs between initial updoming and active rifting, (c) the lateral distribution of erosional thinning across the updomed region, and (d) the post-rifting erosion rate. The longer the time between initial updoming and active rifting, the greater the pre-rifting and post-rifting erosion rates, the thinner the initial crustal thickness, the shorter will be the delay between rifting and marine flooding of the erosionally thinned zone of the continental margin, the continental shelf to be.

A further aspect of crustal thinning concerns the nature of the extreme attenuation that culminates in rupture of the continental crust. The earliest stages can be observed in active continental rift valleys and a slightly later stage has recently been described by Black, Morton, and Varet (1972) for the Afar region of Ethiopia. This early-stage thinning is expressed as normal faulting in the upper brittle crust and perhaps as creep at deeper crustal levels. Although rifted continental margins are very extensive, very little margin is available for reasonably direct inspection. Margins older than about 50 MY have entirely subsided below sea level and are usually masked by sediments. Even the relatively young continental margins of the Gulf of Aden (15–20 MY) are partly submerged and sediment covered. The Somalia shore of the Gulf of Aden (Fig. 4) is nevertheless an excellent area in which to examine a fairly young rifted margin. There is no continental shelf; the margin of the continent forms a 4 km high scarp reaching 1–2 km above sea level and falling abruptly within 60–80 km to 2 km below sea level. This I interpret as a continental margin, still updomed over lithosphere of low density (the rift bulge), and destined to subside below sea level 30–40 MY hence. The 4-km scarp will then be the continental slope, and the Somali Plateau, to the south, will be the continental shelf.

The above sea level portion of the future continental slope is down faulted in subparallel blocks to the north (Beydoun, 1970). Glomar Challenger, on Leg 24 of the Deep Sea Drilling Project, drilled Site 231 (Fig. 4) in 2140 m of water, and penetrated 570 m of hemipelagic sediment overlying ocean floor basalt (Kinsman, in preparation). If

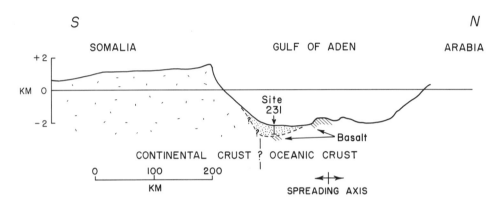

Figure 4. Topographic-bathmetric profile at 48° 10′ East extending from the Somalia plate on the south across the Gulf of Aden, to the Arabian plate on the north. Note the updoming along the continental margin and the steep northward face of the Somalia continental margin, both above and below sea level (this 4-km slope will be the submarine continental slope in 40–50 MY). No continental shelf is presently developed; in 40–50 MY the Somalia plateau will become the submerged continental shelf. The axis of sea floor spreading and the continental-oceanic crust boundary are taken from Laughton et al. (1970). Glomar Challenger, Leg 24, Deep Sea Drilling Project Site 231 is also marked; basalt was penetrated at this site beneath 570 m of hemipelagic sediments. Zone of attenuation or drastic continental crustal thinning is shown to be 60–80 km broad. Note the shape of the uplifted margin; a future basement ridge is already developed at the outer continental terrace-upper continental slope position.

we also use the magnetic anomaly data for the area (Laughton et al., 1970) to help delimit the junction between continental and oceanic crust, a best guess would locate it about 20–25 km landward of Site 231. This does not conflict with other age and sea floor spreading data for the Gulf of Aden. The horizontal distance over which the continent is attenuated from 25–30 km to zero km in thickness is 60–80 km.

This 60–80 km wide belt of attenuation is developed along a segment of rifted continental margin where the separation has comprised a purely divergent motion. In other areas, rifting has been achieved by transcurrent motion; for example, transform faults in the central Gulf of Aden have been traced by Laughton et al. (1970) into the continental margin. Bathymetric profiles across segments of rifted margin where the separating motion was transcurrent almost without exception show much steeper gradients than do profiles of margins where the separating motion was more purely divergent. The evidence suggests that transcurrent rifting succeeds in terminating continents rather more abruptly than does divergent rifting. Along transcurrently rifted margins, the zone of major continental attenuation is only 30–40 km in width. Variations in the width of the zone of major continental margin attenuation will later be reflected as variations in width and thickness of sedimentary accumulations along these margins.

Along most older rifted continental margins, we have only two types of data to tell us something about the horizontal extent over which the major attenuation occurs. The bathymetry itself, of all rifted margins except those built out over by delta deposits, shows the familiar shelf, slope, rise, and abyssal plain profile. The steepness of the continental slope suggests that the major attenuation is relatively abrupt, even though the slope profiles we now see are flattened because of a sediment cover. Intriguingly enough, even fairly mature, sediment mantled, rifted continental margins may still show bathymetric profile differences between segments of margin of divergent and transcurrent origin. The Guinea coast of Africa and the Durban coast of South Africa both have extremely steep continental slopes and are of predominantly transcurrent origin. The south Atlantic margins show both gently sloping divergent and steeper transcurrent segments, as discussed by Francheteau and Le Pichon (1972).

Evidence from seismic refraction and gravity surveys across continental margins (for example, Worzel, 1965; Leyden et al., 1971) suggest that continental crust is generally attenuated rather abruptly, most commonly within horizontal distances of 40–80 km. In a series of papers, Lowell (Lowell and Genik, 1972; Lowell, this volume) has suggested that the Red Sea margins have been attenuated over a very broad zone (150–200 km), but this view is not shared by other authors (for example, McKenzie et al., 1970; Baker et al., 1972) or myself. There is as yet no factual support for models in which the attenuation zone of the continental margin exceeds a width of 80 km.

TECTONIC SYNTHESIS AND A REFINED MODEL
OF CONTINENTAL RIFTING

It seems appropriate now to refine some elements of the previous discussions and then to develop a general model of continental rifting, which will afford constraints on the sedimentary history of rifted continental margins or, conversely, can be tested through analysis of what little we do know of the sedimentary history of these margins.

Rifted continental margins comprise segments that have originated from either a divergent or a transcurrent motion during their early history. Relatively long lengths of predominantly one type or the other may exist. For example, the south Atlantic margins have a predominantly divergent origin for a length of 3000–4000 km, although short lengths of transcurrent margin are interspersed. In contrast, the Guinea coast of west Africa and the northern coast of Brazil have a largely transcurrent origin, for a distance of over 1000 km, although subordinate divergent origin segments are present. Several pieces of evidence suggest that the early histories of these two contrasting types of continental margin are quite different.

Reyment and Tait (1972) have shown how, during the early opening of the south Atlantic, the Guinea coast–northern Brazil coast region remained near sea level, while active transcurrent motion was occurring at the rate of several cm/yr. This conclusion is drawn from the fact that epicontinental marine sediments were deposited over the region during the rifting interval. Minor segments of divergent margin were also involved in the rifting. The lack of continental uplift and the abruptness of the zone of attenuation along transcurrent continental margins suggest that the lithosphere "sheared" and that its thermal structure was little changed during the transcurrent rifting episode. The continental margins were not uplifted, and consequently there was no opportunity for supracrustal erosional thinning to take place. Thus, along mature (>100 MY) transcurrently rifted continental margins we should not find extensive continental shelves or extensive coastal plain sediment wedges; this predicted negative association is confirmed by examination of charts and geological maps.

For divergently rifted continental margins, we have until now used a generalized model of uplift and subsidence based on lithosphere formed at a typical midocean ridge site. We have assumed the subsidence to be related to a time-dependent cooling of the lithospheric slab, all lithosphere having a similar initial temperature and subsequent cooling history. Our assumptions have also considered the asthenosphere to be laterally homogeneous. If these assumptions were all true, we should see linear uplifts preceding continental break-up and comparable erosional thinning along the entire uplifted welt, except for thinning differences related to variations in initial crustal thickness and rock type. The end result after subsidence, should be a belt of sediments along the whole margin that show little variation in width or sediment thickness. What we actually find, however, are domal uplifts along axes of continental rifting and considerable variations in both width and thickness of sedimentary accumulations on the subsided continental crust of mature rifted continental margins. This suggests that our previous discussions of continental break-up have been too simple and that we need to consider reasons for the variation of continental uplift, thinning, and sediment accumulation.

Consideration of a variety of phenomena in the oceans led Wilson (1963, 1965) to hypothesize the presence of hot-spots in the mantle. Morgan (1971, 1972, 1973) extended this idea and was led to suggest the presence of upwardly convecting, cylindrical plumes of deep mantle material, which impinged on the lower surfaces of the lithospheric plates; he suggested that the horizontal viscous flow, spreading radially outward from the plumes at the asthenosphere-lithosphere interface, provided a major driving force for plate motion. The plume pipes were suggested to have a diameter of 100–200 km and their surface manifestation is as topographic highs roughly 1000 km in diameter. Their sites are also elevated (1–2 km) relative to the general depth of the midocean ridge which normally extends between plumes. The igneous products associated with some plumes reach sea level and form volcanic islands; the rocks of these oceanic islands typically have alkali affinities in contrast to the tholeiitic affinities of the basalts of the remainder of the ridge province. Plumes are mostly, but not always, found astride midocean ridges, and the excess volcanism associated with their activity is evidenced in part by the fact that their elevation is greater than the adjacent sea floors. On older sea floors, the trace of a plume is seen as a chain of islands or guyots or as an aseismic ridge, with thicker oceanic crustal development than that found over most of the ocean floor (Bott et al., 1971). Ocean floor evidence suggests that individual plumes may have a life in excess of 10^8 years.

In addition to the major plume sites identified by Morgan (op. cit.), there exists, dotted over the ocean floor, a large number of individual guyots, apparently the product of an almost random distribution of centrally effusive upper mantle activity, but of a smaller scale and of a much more limited duration than the large convective mantle plumes. The recognition of discrete centers of convective upwelling, which bring about thermal and magmatic piercement of the lithosphere, is an important addition to the linear convective upwelling model envisaged for midocean ridges. Lateral inhomogeneities in oceanic crustal thickness, in thermal and possibly petrological character of the oceanic lithosphere and underlying asthenosphere, thus exist.

Le Bas (1971) noted a general correlation on the continents between per-alkaline volcanism, crustal swelling, and rifting, but he considered the orientation of the rifting to be related to older crustal lineaments and that this magmatic and tectonic association was in no way comparable to or related to ocean floor rifting. In other words, continental rifting in this association was not a prelude to sea floor spreading and the formation of a new ocean. He considered that, when plate rupture happened to cut through continental areas, the line of rupture passively followed older lineaments, which might well have been independently followed by an unrelated, earlier episode of continental updoming and rift valley formation. Le Bas demonstrated the existence of domes and associated per-alkaline igneous rocks along the south Atlantic prerift coasts of Africa and South America and similar Tertiary associations in east Africa. He showed that the domes were typically 500–1000 km in diameter, were elevated 1 km or more and that the igneous rocks followed the major updoming period.

I propose to reinterpret the data of Le Bas and suggest that the discrete continental domes are the surface expression of mantle plumes or hot spots and that the continental per-alkaline and alkali igneous rocks associated with the updoming are analogous to the alkali basalts of oceanic plume sites. Morgan (1972) has interpreted the Mesozoic basaltic rocks of the South American and African margins of the south Atlantic as being related to the activity of plumes which became active before the midocean ridge was formed. The presence of a mantle plume beneath a continental area may be caused by either (a) drifting of the plate over a pre-existing plume or (b) inception of a new plume. As most plumes are apparently astride midocean ridges (Morgan, 1972), the first proposal seems unlikely to happen very frequently. Duncan et al. (1972) have proposed that a plume is presently located beneath the

Eifel region in western Europe and that it has apparently been active over the past 35 MY, during which the European plate has moved over it a total of 700 km; it is not certain, however, that the inception of the plume occurred 35 MY ago, as it is possible that its earlier history may be obscured by a cover of younger sediments or by tectonic deformation. During the past 35 MY, passage of the European plate over this plume has left behind a variety of igneous evidences and an up-doming of the continental crust, which has apparently decayed with time. However, no crustal rupture or graben structures seem to be related to this single plume. But when more than one plume is initiated beneath continental crust, the evidence suggests that rifting and graben development rapidly follow. The east African and Ethiopian domes (Baker et al., 1972) were mainly developed over a 15–20-MY period, and their evolution was essentially synchronous.

If one plume was initiated significantly earlier than the others, then we should expect to see an elongate plume trace of igneous activity (cf. the Eifel plume) across one of the continents, marking the plate motion that occurred over the mantle plume, before the other plumes were initiated and continental rifting took place. If the several plumes were coeval, yet did not rift the continent for an extended period of time, we should expect to see an asymmetry to all the domes and their associated igneous manifestations, related to the direction of plate motion over the mantle (if the delay was 20 MY and the plate motion over the mantle was 2 cm/yr, then the originally circular domes would now be roughly ellipsoidal with long axes 400 km greater than short axes). Such plume traces or dome asymmetries are not obviously developed for the south Atlantic domes identified by Le Bas (1971), which suggests either that continental rifting followed fairly rapidly upon plume inception or that there was little motion of the African plate relative to the plume.

The dating of igneous rocks from southern Brazil by Amaral et al. (1967) has demonstrated an age of 120–140 MY for the flood basalts and diabases and for some of the alkaline rocks. This age is close to the time of initial south Atlantic rifting, as determined from sedimentological and paleontological studies (Reyment and Tait, 1972). The age of the greater part of the alkaline igneous rocks, however, is 60–80 MY; the genetic and temporal associations of these alkaline rocks with the Tristan da Cunha plume site are thus uncertain at this time.

Oceanic areas underlain by probable mantle plumes have elevations in excess of those for ridge crests in general, and it seems doubtful that this excess elevation can be accounted for entirely by the presence of a thicker than average oceanic crust. The alternative is to postulate a greater subcrustal mass deficiency than that normally found along ridge crests (Bott et al., 1971). The deeper source of plume materials would suggest that temperatures should be higher at plume sites than at normal ridge sites and the density decrease related to this high temperature would be the most likely cause of the required mass deficiency. If oceanic crust at plume sites is formed at sea level and is 10 km thick and isostatically compensated, we can predict that a 30-km thick continental crust underlain by a similar plume would lie at about 2.6 km above sea level, rather than the value of +1.7 km derived from our normal ocean ridge model (Fig. 2). Continental elevations at updomed plume sites in east Africa and Ethiopia are typically in the range +2–3 km rather than +1–2 km, and elevations fall progressively away from the dome center, as required by the radial flow path of cooling, upwelled mantle plume material. This contrasting thermal structure underlying continental margins of plume versus interplume origin should be reflected in a contrasting subsidence history, the margin updomed over a plume having a higher elevation at all times but subsiding at a faster rate than interplume margins. Bott et al. (1971), in a study of the Iceland–Faroes aseismic ridge and adjacent ocean floor, have demonstrated that similar subcrustal velocities may be measured beneath the two contrasting areas, where the oceanic crust is more than 20 MY in age. These data suggest that after 20 MY the thermal structure of lithosphere of plume and interplume ridge origin is essentially similar and will thereafter follow similar subsidence/time curves.

The smaller scale, central thermal, and magmatic piercement events that lead in the oceanic areas to guyot formation cannot be differentiated in their late thermal and subsidence history from average ocean floor formed at midocean ridges.

In summary it may be possible to distinguish three types of rifted continental margin (Fig. 5): (a) divergent margins updomed initially over a mantle plume, (b) divergent margins of interplume positions, and (c) transcurrent margins. It seems unlikely that a continental margin will be uplifted in the form of a simple elongate welt, to later subside and accumulate a simple linear belt of sediments.

A general model of continental rifting can now be synthesized. Inception of two or more mantle plumes leads initially to discrete updoming of the continent and, within 20 MY, to rupture along a line joining

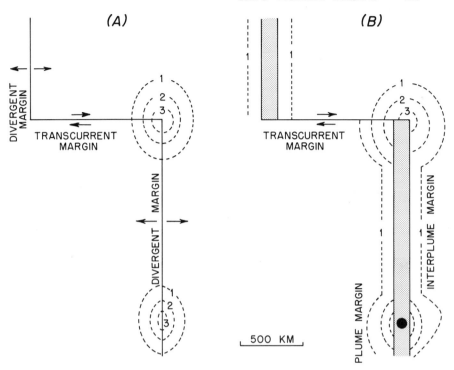

Figure 5. (A) Earliest continental rifting stage, with lengths of divergent and transcurrent margin indicated. Two plume sites are shown where continental crust has been updomed, elevations being marked in km above sea level. Domes are cut by rift valleys. (B) Later rifting period, interplume margins now uplifted in addition to plume margins. Transcurrent margins remain low. New ocean floor stippled; oceanic island over plume site in black, representing a potential "stopper" at the lower end of the juvenile ocean, in which evaporites may thus be deposited.

the plume sites. Rift valley development, volcanism, degree of supra-crustal thinning, attenuation of ruptured continental crust, and later subsidence and sedimentation history will vary along the rifted margin as a direct function of whether a site is located at a plume margin, interplume margin, or transcurrent margin.

Plume Margins

Early domal uplift of continental crust takes place over a mantle plume, to elevations of 2–3 km above sea level and with a dome diameter of 500–1000 km. Rift valleys typically 40 km wide and bounded by large marginal faults are developed across the dome in one or more subparallel chains, and volcanic activity, of considerable intensity, may ensue. Extensive subsidence of rift valley floors and development of lakes are typical. Supracrustal erosional thinning will be fairly intense as this type of margin has the greatest elevation and relief developed. Erosional thinning will extend over the dome, being greatest at the crestal area and possibly totaling 5 to 6 km. After rifting and subsidence, the dome wili appear on each continental margin as a semi-circular or semi-ellipsoidal sedimentary basin of 250–500-km radius, and with a depocenter coinciding with the earlier dome crest area. Subsidence rates during the initial 20 MY after rifting will be somewhat greater than those shown in Figure 2, but thereafter will be similar to those of inter-plume margins. The erosionally thinned rifted continental margin will become submerged 30–60 MY after final continental rupture.

Interplume Margins

Continental margin uplift occurs later than at plume sites and starts when earliest lithospheric attenuation begins; as subcrustal isotherms rise along the potential rifting site, uplift ensues, reaching a maximum of 1–2 km. The horizontal extent of the uplifted welt will be partly dependent on the time that elapses between initial attenuation and active rifting, but 200–300 km is suggested. Erosional thinning is more limited than at plume margins. Figure 2 shows subsidence rates. Rift valley development seems to be more restricted, monoclinal margins are commonly developed; the subsidence of the rift floors is probably less than at plume sites and volcanogenic rocks and sediments may be less well represented than at plume uplifts and associated rift sites. The lacustrine environment also seems less well developed at interplume sites of rifting.

Transcurrent Margins

Little or no uplift occurs and thus no accelerated erosional thinning, no graben development, and probably little or no volcanic activity.

The end result along an extensive rifted continental margin 100 MY after rifting would be an irregularly thinned and irregularly subsided margin. Along transcurrent margins, the narrowness of the attenuation zone would lead to development of very narrow sedimentary basins; the lack of post-rifting subsidence along these margins would be reflected in the lack of a coastal plain sediment wedge. Divergent margins would have wider sedimentary accumulations along their zone of attenuation, but broken into separate half-basins by the location of short lengths of abruptly attenuated transcurrent margin. Variable post-rifting subsidence would lead to formation of extensive coastal plain wedges or sedimentary half-basins at earlier plume sites and less extensive marginal basins at interplume sites.

Sediment Density and the Thickness of Sediment Accumulations

Except for the rift valley environment; ancient continental margin sediments are largely marine. Sedimentation ceases at any locale when the sediment-water interface reaches sea level. The maximum permitted thickness of a pile of sediments is a function of the mean sediment density and of the isostatic loading capacity of the surface on which loading is occurring.

Density-depth data for a wide variety of sedimentary rock types have been compiled by Woollard (1962). Woollard's density-depth envelope for progressively compacted Atlantic and Gulf Coast clayey sediments is indicated in Figure 6. Maxwell (1964) has presented data for quartz sandstones, showing a linear porosity-depth relationship; the data extend to 8-km depth and suggest near-zero porosity at around 10 km. These data provide the low density-depth boundary of Figure 6. The high density boundary of Figure 6 represents well-cemented sandstones and carbonates of low porosity; however, such dense, shallow rocks can generally be shown to have been much more deeply buried at some earlier time in their history. Some individual mineral densities are also included in Figure 6. There are no data for sediment below 10 km, the depth of our deepest hole, but the densities of quartz, feldspar, and calcite— the most abundant minerals of sedimentary rocks—place a strong upper bound on the densities of deeply buried sedimentary rocks. A depth-density profile, used in the subsequent sediment loading discussion, is indicated by the heavy dashed line of Figure 6. The generally rather high porosities of all young, actively accreting sediments and the relatively slow rate at which pore space becomes occluded by cementing minerals when sea water occupies pore spaces suggest that sediment densities within actively subsiding sediment columns will rarely be much higher than the value selected in Figure 6.

Figure 1 shows lithospheric density distributions for continental and

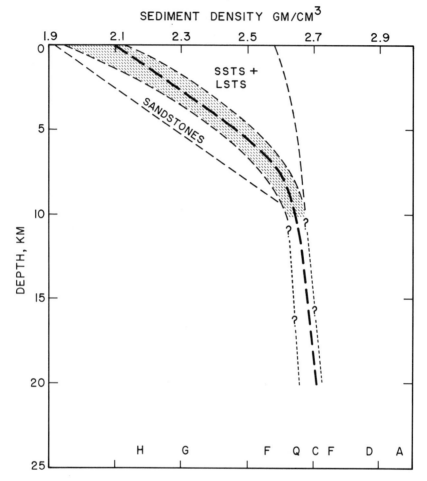

Figure 6. Sediment Density-depth relationships. The stippled envelope is for clayey sediments, both higher and lower density-depth areas being occupied by sandstones (ssts) and limestones (lsts). Projection deeper than 10 km is based on limit defined by quartz (Q), feldspar (F-F), and calcite (C) densities. Other individual mineral densities are halite (H), gypsum (G), dolomite (D) and anhydrite (A). Heavy dashed line is sediment density-depth relationship used in sediment loading discussions.

oceanic columns of 100 MY in age. If we combine these data with the density-depth relationship for sedimentary rocks of Figure 6, it is possible to define maximum sediment loading of a variety of crustal columns. In our crustal model, 30-km thick continental crust overlying 100 MY lithosphere lies close to sea level, and thus the loading capacity of the upper surface by marine sediments is zero. Continental crust less than 30 km thick, overlying 100-MY lithosphere, lies below sea level and marine sediments can therefore accumulate. Figure 7 shows the amounts of sediment that can accumulate on crustal columns of various thicknesses, and, as an example, the sediment density-depth curve from Figure 6 is shown.

Our model also shows 100-MY oceanic crust to be isostatically compensated when loaded with 5.7 km of sea water. Loading with sediment will progressively depress the ocean floor and the loading capacity will finally be taken up when ~17 km of sediments have accumulated. At this depth, the mean sediment density of the entire sediment column is ~2.54 gm/cm^3. This mean density is considerably less than the average quartz-feldspar-calcite density of ~2.7 gm/cm^3. The mean density of any sediment pile is thus seen to be rather sensitive to the very low density/high porosity sediments that typically constitute the upper few km of the section.

The loading capacity of the ocean floor and of thinned rifted continental margin areas is obviously related to subcrustal lithospheric density, and so will change with time, being least on young ocean floor and along young rifted margins and greatest on mature (>100 MY) ocean floor and along old rifted continental margins. The maximum sediment loading of ocean floor of various ages is shown in Figure 8, again using the sediment density-depth curve from Figure 6. On zero-age ocean floor, only 5.2 km of sediment can be deposited before the sediment column reaches sea level, whereas, at 20 MY, ~11 km can accumulate; at 100 MY, ~ 17 km, and, at 150–200 MY, just over 18 km of sediment can be piled up on ocean floor. During the initial 10–15-MY history of a juvenile ocean, salt formation may proceed rapidly enough to take up the loading capacity of the ocean floor; the curve of Figure 8 suggests that 7–8 km of salt could accumulate. Similar loading curves can be constructed for the rifted continental margin itself, and they of course would show similar increases in maximum sediment loading with elapse of post-rifting time.

How do the predicted thicknesses of sediment compare with those

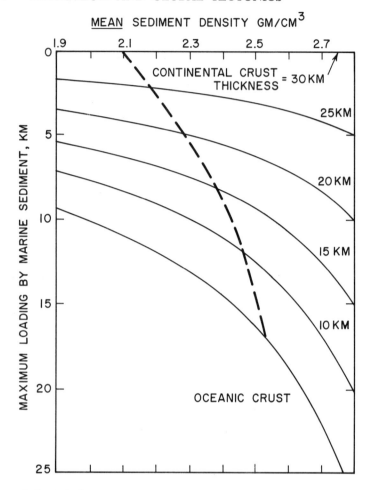

Figure 7. Isostatic loading potential of oceanic crust and continental crust of 30, 25, 20, 15, and 10 km thickness. Mean sediment density curve is calculated from sediment density-depth curve indicated in Figure 6. All crustal columns are underlain by lithosphere ~100 MY in age. Maximum loading of 100 MY ocean floor is close to 17 km.

actually found? The great deltas of the world such as the Mississippi, Niger, Indus, and Ganges have stepped completely off the continent and its thinned margin and are now building out over the ocean floor and burying sea floor topography such as volcanic seamounts. These are the

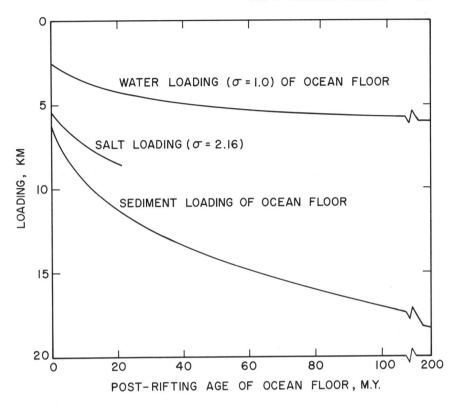

Figure 8. Sediment loading potential of ocean floor with post rifting time lapse, based on ocean floor profile of Figure 2 and sediment density date of Figure 6. The water loaded profile from Figure 2 is included for comparison. The maximum thickness of halite deposited on ocean floor during the first 20 MY of post rifting history is also shown.

obvious locations to examine the predicted maximum loading capacity of ocean floor, and all four are building onto ocean floor that is older than 90 MY (i.e., near its full loading capacity). Moore et al. (1971) have presented data indicating 16.5 km of sediment beneath the outer shelf break of the Ganges cone (n.b., this shelf break is the prograding edge of the delta and not the outer margin of the normal continental terrace). The Niger Delta front is underlain by at least 11 km of sediments (Burke et al., 1971). The Mississippi Delta is underlain by ~15 km of sediments (Murray, 1960). These thicknesses are to be compared with the reconstructed sediment thicknesses of eugeosynclinal sequences in orogenic belts (Stoneley, 1969). Maximum values of 15–17 km often result from such an exercise. These data suggest that the sediment loading model used in this paper is reasonably close to reality.

Continental crustal thicknesses of 25–30 km are typically developed along rifted continental margins where initial updoming permitted supracrustal erosional thinning to proceed, the thinning being greatest toward the outer continental margin. Predicted sediment thicknesses on these marginal areas would be 2–3 km. This is a typical maximum thickness of a shelf and coastal plain sediment wedge, as recorded, for example, along the east coast of the United States (Maher, 1970). Continental crustal thicknesses between 25 and 5 km are developed over the 30–80-km wide attenuated continental margin zone, and here the sediment loading capacity increases rapidly seaward. This is supported by the data of Maher (1971) for example, for the eastern United States continental margin.

Sediment Source Areas and Sediment Supply to Rift Valleys and Rift Oceans

During initial updoming and early rifting, the topographic gradient of the continental surface is directed away from the rift valleys and juvenile ocean, and sediment eroded from the uplifted areas will be largely transported away from these sites. Topographic maps show sediment source areas of active rift valleys and also of juvenile oceans (<20–30 MY) to be 50–100-km wide, but relief is high, erosion rates are high and considerable volumes of sediment can be derived from even these rather limited source areas. Certainly sediment supply can normally keep up with rates of rift valley basin subsidence, although at times, for reasons that are not at all obvious, sediment starvation conditions can prevail (cf. the present-day Lake Tanganyika rift valley basin).

As the juvenile ocean widens, the drainage divide along the continental margin will slowly retreat by erosion; in addition, the gradual subsidence of the rifted continental margin will also tend to move the drainage divide landward. After about 80–100 MY, the entire continental area may drain toward the trailing margin, as, by that time, the leading edge of the continent has been thickened and uplifted, and the continental drainage divide will be located somewhere near the leading margin. This situation is well portrayed by South America, the drainage of which now largely enters the ocean off the trailing margins along the Atlantic. However, as drainage area increases, so relief decreases, and as relief decreases the detrital to solution ratio of denudation products decreases. The result is that sediment supply to the rifted margin does not increase in proportion to the source area.

A sediment source area 100 km wide would need to be eroded at a rate of $250 \text{ mm}/10^3$ yr in order to supply a volume of detrital sediment approximately equal to the annual new volume of ocean formed, if sea floor spreading proceeds at a half-rate of 1 cm/yr. Such high erosion rates are found today only in the Himalayas and a few other extreme situations and would seem unreasonably high for a juvenile ocean margin. The conclusion, then, is that, if sea floor spreading proceeds at a half-rate in excess of 1 cm/yr, an ocean is assured, as sediment infilling cannot match the rate at which the ocean increases in volume.

SEDIMENTARY ACCUMULATIONS OF RIFTED CONTINENTAL MARGINS

Rift Valley Basins

Rift valley basins are as yet poorly and inadequately studied. We know more about ancient, presumed rift valleys than we do about modern ones. One or several subparallel rift valleys are typically formed early in the history of continental rifting along a divergent margin. In their early history, rift valleys are often structurally defined by marginal monoclines; at a later stage, steep (65°), normal faults develop, extend to at least 10 km in depth, and are typically 30–50 km apart (Baker et al., 1972). A single rift belt is often transversely segmented by large faults, and graben blocks themselves may exhibit differential vertical motion in excess of 5 km. Differential subsidence between the elevated rift margins and axial graben blocks commonly exceeds 5 km and may reach nearly 10 km. Rates of subsidence of rift valley floors may exceed 0.5 km/MY. Sediment source areas are small, but erosion rates are high and sediment supply may keep up with rates of subsidence. During such periods, fluviatile facies may be developed over the entire basin. However, a state of sediment starvation is commonly developed, leading to formation of extensive, often very deep (>1 km) rift valley lakes, the floors of which may lie below sea level, as in the present Lake Tanganyika. Marginal fanglomerates and shoreline sands pass laterally into fine-grained lacustrine sediments, and sometimes organic oozes, such as diatomites, are well developed. Turbidites have been recorded in the deeper basins of Lake Tanganyika by Degens et al. (1971). The deltaic facies may be well developed, typically at one end of a lake with axial drainage, for example the Omo Delta of present day Lake Rudolf (Butzer, 1971). Rift valley lakes are commonly nutrient-rich and stratified; anoxic bottom waters, combined with high organic productivity rates and slow detrital sediment input rates may result in a lacustrine facies with a high organic fraction (5–10% in some modern Lake Tanganyika sediments; Degens et al., 1971).

All ancient rift valleys that have been described, such as the Triassic basins of the eastern United States, the Reconcavo-Tucano basin of Brazil, and several others, are sites where continental rifting underwent conception, followed by a short-term gestation, but the rifting was finally aborted. These rift valley grabens and their marginal faults have subsequently annealed and become part of the continent once again. Where are the rift valleys that developed along the suture where rifting did ultimately occur? We must surely look for these thick (5–10 km), narrow (40 km), sedimentary masses at the outermost edge of the rifted continental margin, perhaps in the zone of continental margin attenuation. The major aspect of the rift valley basin marginal faults leads me to hypothesize that the rift valley basin will most likely be left in its entirety attached to one continental margin or the other, but that it is unlikely to be split and fragments left attached to both margins. The laterally segmented structure of the rift valley province may lead to a complex rifting pattern, discrete lengths of rift basin being left attached to one margin or the other, so that the outermost continental margin will show a discontinuous occurrence of 40 km wide, 5–10 km thick sedimentary masses. Where these are absent, the margin will terminate in fractured basement rocks.

Sedimentation Against the Attenuated Continental Margin and on the Floor of the Juvenile Ocean

If much of the continental margin attenuation is accomplished after the intracontinental rift valley stage, then the 30–80 km extent of this zone, at a horizontal spreading half rate of 1–2 cm/yr, could take 1.5–8 MY to develop. Complete rupture of the continent would then have been achieved, and the sea floor, for example at interplume sites, would be formed at depths of 2–3 km below sea level. The continental margin would at that time comprise a 2–3 km section below sea level and a 1–2 km section above sea level (Fig. 4). The upper surface of the 2–3-km below-sea level section will probably have been at sea level during the attenuation stage and should be mantled with nonmarine sediments, overlain progressively by shoreline and shallow marine facies. This transgressive sequence will gradually move upward over the upper 1–2-km section of the attenuated continental margin during the 30–50 MY post-rifting subsidence period, as cooling of the subcrustal lithosphere permits the continental margin to subside. Finally, 30–50 MY after rifting, marine facies sediments will transgress the continental terrace, forming a coastal plain wedge and shelf sequence. The first sediments formed on ocean floor will be marine facies, deposited in 2–3 km of water if the ocean floor was formed at an interplume site or at shallower depths on plume-related sea floor. Thus the oldest marine sediments are laid down on the lower continental slope; the oldest marine sediments deposited on ocean floor will be 2–8 MY younger; the oldest marine sediments deposited on the upper continental terrace will be 30–50 MY younger than those formed on the oldest ocean floor (Figures 9 and 10).

Evaporite Deposits of Rifted Continental Margins
and Juvenile Oceans

Minor nonmarine evaporites are formed in rift valley basin sequences. Formation of marine evaporites requires extreme restriction of sea water circulation. This is evidenced today by the Mediterranean and Red Sea, neither of which is a brine basin in spite of almost complete isolation from the world ocean. The emphasis on restriction of circulation of diluting waters is dictated by the rather limited evaporation that can be achieved in natural environments. Effective barriers to circulation must extend above sea level and comprise either continental crustal areas or, more rarely, oceanic islands. Rift valleys that intersect an older continental margin or juvenile oceans with elevated marginal continents are ideal environments for evaporite formation. For example, during the opening of the south Atlantic, rifting along the Guinea Coast was largely transcurrent, and the juvenile ocean was blocked at its northern end. At its southern end, the Tristan da Cunha plume site probably formed an oceanic island represented today by the landward ends of the Rio Grande Rise and Walvis Ridge. This oceanic island effectively restricted sea water circulation into the ocean to the north, net evaporation prevailed, and evaporites were deposited, both against the attenuated continental margin itself and on the first formed ocean floor. The

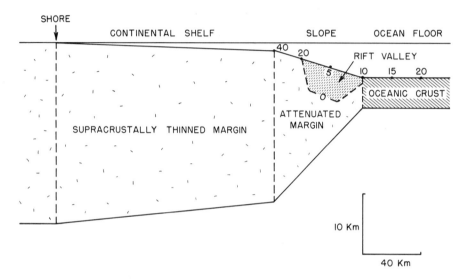

Figure 9. Divergently rifted continental margin 100 MY after rifting. Loading by water only, except for a rift valley basin (stippled) in the attenuated margin zone. Numbers represent MY after rift valley inception; 0–5 MY, rift valley infilling; 5–10 MY, continental margin attenuation period; between 5 and 10 MY the first marine sediments are laid down on the outer part of the attenuated continental margin. At 15–20 MY the margin profile is similar to Figure 4. At 40 MY or as late as 60 MY, the first marine sediments are deposited on the continental terrace.
All faults and structural elements of scale smaller than 10 km are omitted; although drawn to scale, the figure should be considered a cartoon.

evaporites of the African margin immediately north of the Walvis Ridge are calcium sulphate deposits, whereas those farther north along the margin are dominated by halite; this is just the pattern to be expected if influx of sea water was over the Rio Grande–Walvis barrier. The age of the evaporites is known to be Aptian in the marginal basins. The width of the zone of diapirs of possible halite along the African margin

(Leyden et al., 1971), which are located on the oldest ocean floor and seaward attenuated continental margin area, is about 200 km, which at a horizontal spreading rate of 2 cm/yr suggests a time period for salt deposition of around 10 MY. The ability of an oceanic island at a plume site to act as an effective stopper to restrict sea water circulation into a juvenile ocean is obviously limited because of time-dependent subsi-

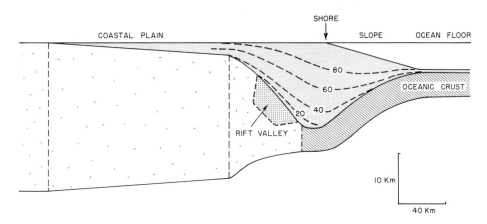

Figure 10. A cartoon, much as Figure 9, but indicating the sediment loading capacity of a fully loaded, mature, divergently rifted, continental margin and adjacent oceanic crust. Numbers refer to evolving sediment pile in millions of years after continental rupture. If evaporites are present they will occur beneath the 20 MY contour, at the base of the maximum thickness of the sediment pile. They will lie on both attenuated continental crust and on oceanic crust. The vertical and horizontal dimensions of the sediment pile across the continental margin are all taken from this paper; note the strong resemblance of this pre-dicted sediment geometry to that known under large deltas such as the Mississippi or Bengal cones.

dence as cooling proceeds and the south Atlantic suggests 10 MY as the effective lifetime of the Tristan da Cunha barrier.

In the Red Sea, the exact form of the Gulf of Suez region, which was the barrier over which sea water reached the evaporite basin, is unknown. However, the present Red Sea configuration at its southern end suggests another effective barrier to sea water entry into juvenile ocean basins, in the form of continental fragments, such as the Danakil Alps. The size of these fragments and the horizontal spreading rate will determine the life span of these evaporite basin barriers. In general the probability of evaporite formation in a juvenile rifted ocean decreases with time. Evaporite formation is most likely to occur during the initial 5–10 MY. The widespread occurrence of rifted continental margin evaporites, many of them several km in thickness (Anonymous, 1965), points to the frequency with which effective sea water circulation barriers are developed during the juvenile ocean stage of rifting. The thickness of some of these rift ocean evaporites is to be noted; in the western part of the southern Red Sea (Hutchinson and Engels, 1972) a profile at right angles to the axial rift shows 3–4 km of evaporites with intercalated basalts just west of the present axial rift and 7–8 km of evaporites near the present Red Sea shoreline. The landward thickening of sediment columns is the pattern predicted for ocean floor, the loading capacity of which increases with increasing age. These evaporite thicknesses are relatively close to those predicted for halite loading of oceanic crust (Figure 8). To date, there is no definite evidence to support suggestions that the Red Sea evaporites are underlain by anything other than basaltic oceanic crust, and certainly the evaporite thickness distributions would support this.

If evaporite deposits are so commonly formed against attenuated continental margins and on the adjacent oldest ocean floor, then they should be deeply buried during the later history of a trailing margin, as sediment eroded from the continent accumulates above them. The evaporites are likely to be buried at the base of a pile of sediment that may be up to 16–17 km in thickness. The occurrence of such evaporites, flooring eugeosynclinal sediment piles, may have possible structural importance during the later compressive history of the margin.

References

Amaral, G., J. Bushee, U. G. Cordani, K. Kawashita, and J. H. Reynolds, 1967. Potassium-argon ages of alkaline rocks from southern Brazil, *Geochim. et Cosmochim.* Acta v. 31, pp. 117–142.

Anonymous, 1965. *Salt Basins around Africa.* Publ Inst. of Petroleum, London, 122 pp.

Baker, B. H., P. A. Mohr, and L. A. J. Williams, 1972. Geology of the eastern rift system of Africa. Geol. Soc. Amer. Spec. Paper 136, 67 pp.

Beydoun, Z. R., 1970. Southern Arabia and northern Somalia; comparative geology, *Phil. Trans.* Roy Soc. London. v. 267, pp. 267–292.

Bird, J. M., R. Isacks, eds., 1972. *Plate Tectonics: Selected Papers from the Journal of Geophysical Research.* Amer. Geophys. Union. 563 pp.

Black, R., W. H. Morton, and J. Varet, 1972. New data on Afar tectonics. *Nature,* v. 240, p. 170–173.

Bloom, A. L., 1967. Pleistocene shorelines; a new test of isostasy. *Geol. Soc. Amer. Bull.,* v. 78, pp. 1477–1494.

Bott, M. H. P., C. W. A. Browitt, and A. P. Stacey. 1971. The deep structure of the Iceland-Faeroe Ridge, *Marine Geophys. Researches,* v. 1, pp. 328–351.

Brotchie, J. F., and R. Silvester, 1969. On crustal flexure, *Jour. Geophys. Res.,* v. 74, pp. 5240–5252.

Burke, K., T. F. J. Dessauvagie, and A. J. Whiteman, 1971. Opening of the Gulf of Guinea and geological history of the Benue Depression and Niger Delta, *Nature,* v. 233, pp. 51–55.

Butzer, K. W., 1971. *Recent History of an Ethiopian Delta.* Univ. of Chicago Res. Paper 136, 184 pp.

Crittenden, M. D., 1963. *New data on the isostatic deformation of Lake Bonneville,* U.S. Geol. Surv. Prof. Paper 454-E, 31 pp.

Degens, E. T., R. P. von Herzen, and H. Wong, 1971. Lake Tanganyika; water chemistry, sediments, geological structure, *Naturwissenschaften,* v. 58, pp. 229–241.

Dietz, R. S. 1963. Collapsing continental rises; an actualistic concept of geosynclines and mountain building, *Jour. Geol.,* v. 71, pp. 314–333.

———, 1965. Collapsing continental rises: an actualistic concept of geosynclines and mountain building: a reply. *Jour. Geol.,* v. 73, pp. 901–906.

———, 1966. Passive Continents, spreading sea floors, and collapsing continental rises, *Amer. Jour. Sci.,* v. 264, pp. 177–193.

Duncan, R. A., N. Petersen, and R. B. Hargraves, 1972. Mantle plumes movement of the European plate, and polar wandering, *Nature,* v. 198, pp. 82–86.

Fischer, A. G., 1973. Origin and growth of basins (this volume).

Fancheteau, J., and X. Le Pichon, 1972. Marginal fracture zones as structural framework of continental margins in South Atlantic Ocean, *Amer. Assoc. Petrol. Geol. Bull.,* v. 56 pp. 991–1007.

Gilluly, J., 1964. Atlantic sediments, erosion rates, and the evolution of the continental shelf: some speculations, *Geol. Soc. Amer. Bull.,* v. 75, pp. 483–492.

Gutenberg, B., 1941. Changes in sea level, postglacial uplift, and mobility of the Earth's interior, *Geol. Soc. Amer. Bull.,* v. 52, pp. 721–772.

Hess, H. H., 1960. Caribbean Research Project: Progress Report. *Geol. Soc. Amer. Bull.,* v. 71, pp. 235–240.

Holmes, A., 1965. *Principles of Physical Geology.* New York, Ronald Press, 1288 pp.

Hospers, J., 1965. Gravity field and structure of the Niger delta, Nigeria, West Africa, *Geol. Soc. Amer. Bull.,* v. 76, pp. 407–422.

Hsü, K., 1958. Isostasy and a theory for the origin of geosynclines, *Amer. Jour. Sci.*, v. 256, pp. 305–327.

———, 1965. Isostasy, crustal thinning, mantle changes, and the disappearance of ancient land masses, *Amer. Jour. Sci.*, v. 263, pp. 97–109.

Hutchinson, R. W., and G. G. Engels, 1972. Tectonic evolution in the southern Red Sea and its possible significance to older rifted continental margins, *Geol. Soc. Amer. Bull*, v. 83, pp. 2989–3002.

Isacks, B., J. Oliver, and L. R. Sykes, 1968. Seismology and the new global tectonics. *Jour. Geophys. Res.*, v. 73, pp. 5855–5899.

Judson, S., and D. F. Ritter, 1964. Rates of regional denudation in the United States, *Jour. Geophys. Res.*, v. 69, pp. 3395–3401.

Kinsman, D. J. J. (in press). Gulf of Aden; summary of Leg 24 drilling results. In: Fisher, R. L., Bunce, E, Kinsman, D. J. J. (and others), *Leg 24, Initial Reports of the Deep Sea Drilling Project, Gulf of Aden and Indian Ocean*. Washington, D.C., GPO.

Kossinna, E., 1921. *Die Tiefen des Weltmeeres*. Berlin Univ. Institut f. Meereskunde, Veröff., Geogr.-Naturwiss, Kl. 9, 70 pp.

Langbein, W. B., and S. A. Schumm, 1958. Yield of sediment in relation to mean annual precipitation, *Trans. Amer. Geophys. Union*, v. 39, pp. 1076–1084.

Laughton, A. S., R. B. Whitmarsh, and M. T. Jones, 1970. The evolution of the Gulf of Aden, *Phil. Trans. Royal Soc. London*, v. 267, pp. 227–266.

Lawson, A. C., 1942. Mississippi Delta—a study in isostasy, *Geol. Soc. Amer. Bull.*, v. 53, pp. 1231–1254.

Le Bas, M. J., 1971. Per-alkaline volcanism, crustal swelling and rifting, *Nature* v. 230, pp. 85–87.

Leyden, R., W. J. Ludwig, and M. Ewing, 1971. Structure of continental margin off Punta del Este, Uruguay and Rio de Janeiro, Brazil. *Amer. Assoc. Petrol. Geol. Bull.*, v. 55, pp. 2161–2173.

Leyden, R., G. Bryan, and M. Ewing, 1972. Geophysical reconnaissance on African shelf: 2. Margin sediments from Gulf of Guinea to Walvis Ridge. *Amer. Assoc. Petrol. Geol. Bull.*, v. 56, pp. 682–693.

Lowell, J. D., et al., 1973. Petroleum and plate tectonics of the southern Red Sea (this volume).

Lowell, J. D., G. J. Genik, 1972. Sea-floor spreading and structural evolution of southern Red Sea, *Amer. Assoc. Petrol. Geol. Bull.*, v. 56, pp. 247–259.

Maher, J. C., 1971. Geological framework and petroleum potential of the Atlantic coastal plain and continental shelf, *U.S. Geol. Surv. Prof. Paper* 659, 98 pp.

MacKenzie, D., 1967. Some remarks on heat flow and gravity anomalies, *Jour. Geophys. Res.*, v. 72, pp. 6261–6273.

MacKenzie, D. P., D. Davies, and P. Molnar, 1970. Plate tectonics of the Red Sea and East Africa, *Nature*, v. 226, pp. 243–248.

MacKenzie, D., and J. G. Sclater, 1971. The evolution of the Indian Ocean since the late Cretaceous, *Geophys. Jour. Roy. Astron. Soc.,* v. 25, pp. 437–528.

Maxwell, J. C., 1964. Influence of depth, temperature and geologic age on porosity of quartzose sandstone, *Amer. Assoc. Petrol Geol. Bull.* v. 48, pp. 697–709.

Meade, R. H., 1969. Errors in using modern streamload data to estimate natural rates of denudation, *Geol. Soc. Amer. Bull.,* v. 80, pp. 1265–1274.

Menard, H. W., 1961. Some rates of regional erosion, *Jour. Geol.,* v. 69, pp. 154–161.

Menard, H. W., and S. M. Smith, 1966. Hypsometry of ocean basin provinces, *Jour. Geophys. Res.,* v. 71, pp. 4305–4325.

Moore, D. G., J. R. Curray, and R. W. Raitt, 1971. Structure and history of the Bengal deep-sea fan and geosyncline, Indian Ocean. Abst. VIII Internat. Sediment. Cong. Heidelberg, p. 69.

Morgan, W. J., 1971. Convection plumes in the lower mantle, *Nature,* v. 230, pp. 42–43.

———, 1972. Deep mantle convection plumes and plate motions, *Amer. Assoc. Petrol. Geol. Bull.,* v. 56, pp. 203–213.

Murray, G. E., 1960. Geologic framework of Gulf Coastal province of United States. In *Recent Sediments, Northwest Gulf of Mexico,* Ed. F. P. Shepard et al., Pub. Amer. Assoc. Petrol. Geol., 394 pp.

Reyment, R. A., and E. A. Tait, 1972. Biostratigraphical dating of the early history of the South Atlantic Ocean, *Phil. Trans. Roy. Soc. London.,* Ser. B, v. 264, pp. 55–95.

Rona, P. A., 1970. Comparison of continental margins of eastern North America at Cape Hatteras and northwestern Africa at Cap Blanc, *Amer. Assoc. Petrol. Geol. Bull.,* v. 54, pp. 129–157.

———, 1972. Relation between rates of sediment accumulation on continental shelves and sea floor spreading in the central North Atlantic, *Geol. Soc. Amer. Abst.,* v. 4, pp. 644.

Schumm, S. A., 1967. The disparity between present rates of denudation and orogeny. U.S. Geol. Surv. Prof. Paper 454-H. 13 p.

Sclater, J. G., and J. Francheteau, 1970. The implication of terrestrial heat flow observations on current tectonic and geochemical models of the crust and upper mantle of the earth, *Geophys. Jour. Royal. Astronom. Soc.,* v. 20, pp. 509–542.

Sleep, N. H., 1971. Thermal effects of the formation of Atlantic continental margins by continental break up, *Geophys. Jour. Roy. Astronom. Soc.,* v. 24, pp. 325–350.

Stonely, R., 1969. Sedimentary thicknesses in orogenic belts. In *Time and Place in Orogeny,* ed. P. E. Kent, G. E. Satterthwaite, A. M. Spencer, Geol. Soc. London Spec. Pub. No. 3.

Ten Brink, N. W., 1972. Glacio-isostasy; new data from West Greenland

and geophysical implications: *Geol. Soc. Amer. Abst.,* v. 4, No. 7, pp. 686–687.

Van Houten, F. B., 1969. Late Triassic Newark Group, north-central New Jersey and adjacent Pennsylvania and New York. In *Geology of Selected Areas in New Jersey and Eastern Pennsylvania and Guidebook of Excursions.,* ed. S. Subitzky, Rutgers University Press, 382 p.

Walcott, R. I., 1970. Flexural rigidity, thickness and viscosity of the lithosphere, *Jour. Geophys. Res.,* v. 75, pp. 3941–3954.

————, 1972. Gravity, flexure, and the growth of sedimentary basins at a continental edge, *Geol. Soc. Amer. Bull.,* v. 83, pp. 1845–1848.

Wilson, J. T., 1963. Hypothesis of earth's behavior, *Nature,* v. 198, pp. 925–929.

————, 1965. Evidence from ocean islands suggesting movement in the earth,, *Phil. Trans. Royal Soc. London,* v. 258, pp. 145–167.

Wollard, G. P., 1962. The relation of gravity anomalies to surface elevation, crustal structure and geology. Technical Report AG-23-601-3455 (62–9), Univ. of Wisconsin, 78 pp.

Worzel, J. L., 1965. Deep structure of coastal margins and midoceanic ridges. In *Colston Symposium,* v. 17, ed. W. F. Whittard and R. Bradshaw, London, Butterworths, 464 pp.

From the more general and theoretical treatment by Kinsman, we now proceed to the specific description and interpretation of the world's classical active rifts basin, the Red Sea. James Lowell and his colleagues of the Esso organization bring to this topic the extensive geophysical and geological experience and data that resulted from petroleum exploration in the Red Sea area under the auspices of the Standard Oil Company of New Jersey and Mobil Oil. The implications of this study extend, of course, to rift basins in general.

Petroleum and Plate Tectonics of the Southern Red Sea

JAMES D. LOWELL,[1]
GERARD J. GENIK,[2]
THOMAS H. NELSON,[3]
AND
PAUL M. TUCKER[4]

ABSTRACT

The Nubian and Arabian plates are being separated in the southern part of the Red Sea by two concurrently spreading rifts. This continental dispersal is a continuing event in a sequence of structural evolution of arching, rifting, subsidence, and breaching of continental lithosphere that has influenced in a profound way virtually every control on the occurrence of petroleum in the area.

During the arching stage (mainly Oligocene), continental sandstones of

[1] Northwest Pipeline Corp., Denver, Colorado 80202. Formerly with Exxon Co., U.S.A.
[2] Esso Exploration, Inc., Walton-on-Thames, England
[3] Esso Production Research Co., Houston, Texas 77001
[4] Esso Production Research Co., Houston, Texas 77001 (Retired)
We are grateful to H. R. Gould, R. Sarmiento, and D. A. White of Esso Production Research Company, C. A. Burk of Mobil Oil Corporation, and the Princeton University Conference directors A. G. Fischer and Sheldon Judson for encouraging us to present and prepare this paper. We thank Hershel Garrett and Janet Pendleton for their masterful preparation of illustrations. Esso Production Research Company, Esso Exploration, Inc., and Mobil Oil Corporation kindly granted permission to publish.

reservoir potential may have formed when short streams drained developing fault blocks and scarps; volcanism also occurred at this time. Such a nonmarine environment seems probable for the interbedded clastics and volcanics of the Trap Series (Oligocene-Miocene) in Yemen and the Dogali Series (latest Eocene-earliest Miocene) in Ethiopia, and also for similar correlative lithologies in other countries peripheral to both the Red Sea and the Gulf of Aden.

The rifting stage (mainly Miocene) led to pronounced subsidence on horst-and-graben normal-fault blocks, one of two dominant structural styles in the Red Sea, and the style that has formed the preponderance of potential structural traps. The second style is that of salt structures derived from a thick salt sequence that was deposited in a narrow, restricted, rifted Miocene trough that formed as a consequence of breakup of continental lithospheric plates. In addition to providing structures, the salt is ideally situated to seal hydrocarbons in underlying porous reservoirs.

During the double rifting stage (Pliocene to present), the continental lithosphere was breached to allow insertion of oceanic crust in the axial rift and was considerably thinned in the rift to the west. Double rifting has led to extremely high heat flows, probably greater than for breakup of a lithospheric plate by single rifting. The southern Red Sea is consequently believed to be gas-prone. This interpretation is at least partially confirmed by a gas blowout drilled in Ethiopian Red Sea waters.

Some of the relationships between petroleum and plate tectonics demonstrated in the southern Red Sea can no doubt be applied to other lithospheric plates of rifted origin.

INTRODUCTION

As partners, Esso Exploration, Inc., and Mobil Oil Corporation explored for petroleum in Ethiopian waters of the Red Sea from 1966 through 1970. It is doubtful if any other exploration venture in history has been so strongly influenced by the manner of breakup and dispersal of continental lithospheric plates with all its ramifications of arching, normal faulting, volcanism, high heat flow, evaporite accumulation, and deposition of potential reservoir rocks. The Red Sea as a site of rifting of continental lithosphere where the global rift system continues into a continent (Figs. 1, 2) has been established by seismicity, gravity, heat flow, magnetics, seismic refraction, and direct sampling; this evidence

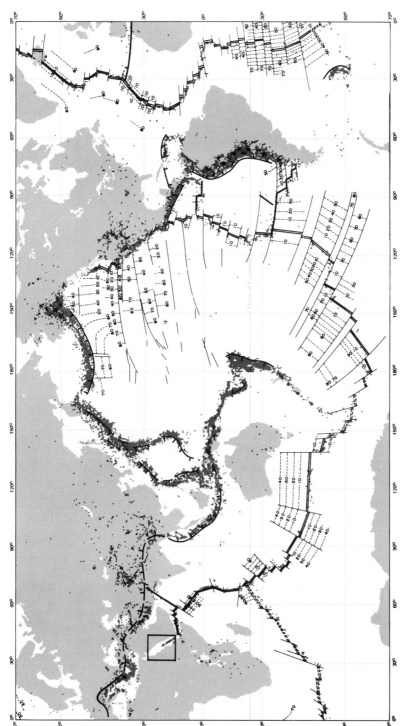

Figure 1. Position of Red Sea in relation to global rift system. Midocean ridges (solid double line, dashed where uncertain) are defined by earthquake epicenters (dots) and isochrons (thin dashed lines) which were derived from magnetic patterns and give age of sea floor in millions of years; ridges are offset by transform faults (thin solid lines). Subduction zones are indicated by heavy hachured line. (From *Sea-Floor Spreading*, by J. R. Heirtzler. Copyright 1968 by Scientific American, Inc. All rights reserved.)

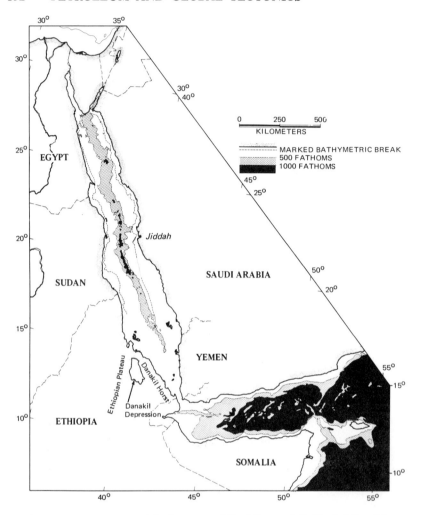

Figure 2. Index map and bathymetry of Red Sea and Gulf of Aden. The early evolutionary stages of rifting and drifting of continental lithospheric plates are reflected in the bathymetry by the Gulf of Aden's having a midocean ridge with crest and flanks, surrounded by small abyssal plains floored by oceanic crust, which in turn are bounded by the incipient continental rise, slope, and shelf both to north and south. A prominent axial rift valley is present along much of the length of the Red Sea, and a midocean ridge has not yet evolved. In approximately the southern two thirds of the Red Sea, the rift valley is floored with oceanic crust, while the entire width of the northern one third is floored by greatly thinned continental crust. Oceanic basalts have not yet broken through. The northern third of the Red Sea is thus in an earlier evolutionary stage than the southern two thirds, which in turn is less mature than the Gulf of Aden as an evolving ocean basin. Bathymetry from Laughton (1970) and Laughton et al. (1970). Courtesy Amer. Ass'n Petroleum Geol.

has been listed by Lowell and Genik (1972). In order to demonstrate the close tie between petroleum and plate motion, the structural evolution and amount of plate separation must be considered, as well as the controls on hydrocarbon occurrences themselves. Inasmuch as Lowell and Genik (1972) have recently published an interpretation of the structural evolution and plate separation, however, these are, with the exception of the section on arching, only summarized here. The main purpose of this paper is to emphasize the influence of structural evolution and plate separation on the controls of petroleum occurrence.

STRUCTURAL EVOLUTION

Description

The structure section of Figure 3 is representative of the present structural configuration of the southern Red Sea. The section is well controlled in the upper few kilometers, but the Moho and base of the lithosphere are drawn by inference. The most striking feature of the section is the thinning of the continental lithosphere that must have taken place to account for the volcanics and oceanic crust in the axial trough. Much of this thinning took place in the brittle portion of the lithosphere by rotation of rigid blocks by listric normal faulting; normal faults are well known from surface geologic and seismic reflection data. Thinning could not have taken place only by normal faulting of brittle material, however, because this would have required extreme rotation (approaching 90°) of beds on normal faults for which no evidence is present from seismic reflection records. Other thinning mechanisms must also have been operative, probably including microscale extensional deformation internal to normal fault blocks and plastic deformation abetted by high temperatures below the brittle limit of the lithosphere. Another salient feature of the Figure 3 section is the great thickness of salt of Miocene age. More than 3 km of salt was drilled by the Amber well, and the reflection seismograph (Fig. 4) suggests that 5 km is present in the deeper parts of the Ethiopian Red Sea basin.

Finally, a second rift is shown to the west of the axial rift. It is spectacularly demonstrated on a reflection seismic line (Fig. 4) by large normal faults that have displaced the sea bottom by as much as 400 m, and basement by more than 2 km. South of the region of the intersec-

		Q	ALLUVIUM (and other unconsolidated material)
QUATERNARY			
TERTIARY	MIO-PLIOCENE	Tde	DESSET SERIES (clastics and coral reefs; unconformity at base)
	MIDDLE-UPPER	Th/Ts	SALT
	MIOCENE		HĂBAB FM. (includes *Globigerina* marls)
	LOWER		
	OLIGOCENE	Td To	DOGALI SERIES (clastics with fish remains; lavas with pillows; tuffs)
CRETACEOUS			
JURASSIC		J	
PERMIAN OR OLDER		Pw	WAJID SANDSTONE
PRECAMBRIAN		PЄ	CRYSTALLINE BASEMENT

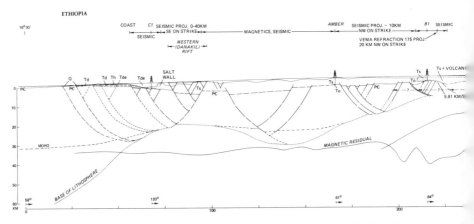

Figure 3. Index map of southern Red Sea, stratigraphic legend, and structure section from northern part of Ethiopian Plateau across southern Red Sea north of Dahlak Islands and through Farasan Islands to southern corner of Saudi Arabia. Sedimentary rocks have been encountered below the Miocene salt (Ts) in the Cl, Amber, and Mansiyah wells, and they are believed to be present from considerations of structural evolution (see text on Structural Evolution—Inter-

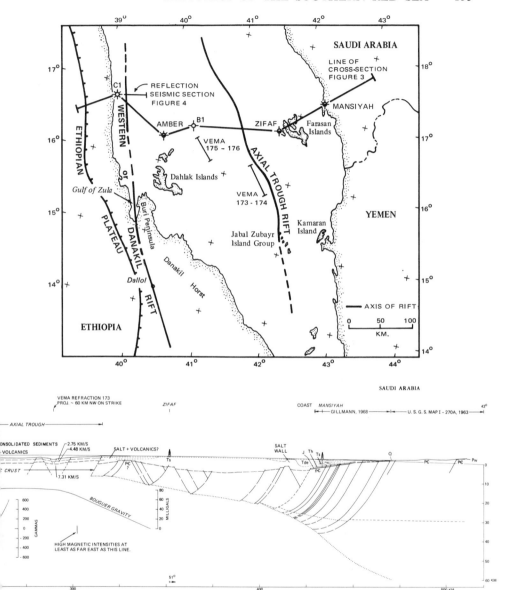

pretation). However, they have never been completely penetrated, and their distribution is virtually unknown. Lacking sufficient control, we have shown salt to rest directly on granitic basement (ᴘᴇ) on the structure section (see also Fig. 4). (Bathymetry from bathymetric map; topography from U.S.A.F. Jet Navigation charts; gravity and, in part, magnetics from Allan, 1970.) Published with permission of the American Association of Petroleum Geologists.

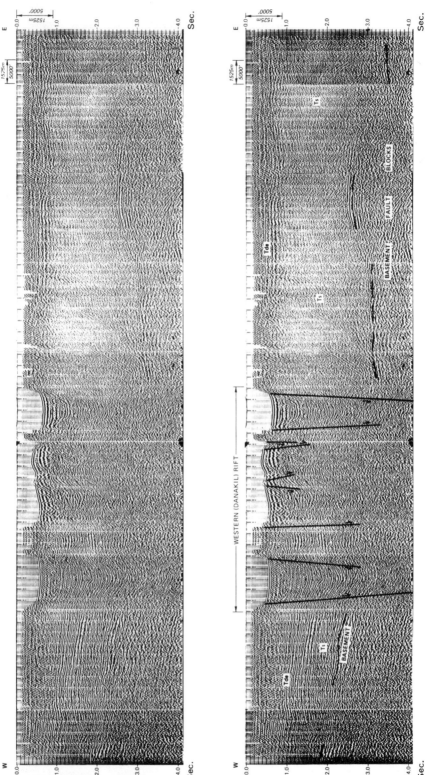

Figure 4. Reflection seismic line in western part of southern Red Sea. (See Fig. 3 for line of section and stratigraphic legend.) Published with permission of the American Association of

tion of the seismic line with the rift is the north-northwest trending Gulf of Zula bounded by a large basement block—the Buri Peninsula on the east and, on the west, by the Red Sea coastal plain, which is down-faulted from the Ethiopian Plateau lying still farther west (Fig. 3). A large down-to-the-west normal fault bounds the northern projection of the Buri Peninsula on its west side (Frazier, 1970), making the Gulf of Zula a large graben. Landward from the Gulf of Zula is the Danakil Depression, a feature currently marked by volcanism, seismicity, and normal faults along the Ethiopian Plateau to the west and the Danakil Horst to the east. The Danakil or western rift and the axial rift of the Red Sea thus constitute two concurrently developing rifts which run nearly parallel some 400 km and possibly much farther.

Interpretation

A summary of the interpretation of the structural evolution of the southern Red Sea is given in Figure 5. Section V of Figure 5 is a simplified reduction of the structure section of Figure 3. Its evolution can be considered in three stages: an arching stage represented by section I; a rifting stage represented by sections II and III; and a double rifting stage represented by sections IV and V.

ARCHING

We believe that the first step in the structural evolution of the southern Red Sea was the regional arching of continental lithosphere with concomitant volcanism, erosion, and sedimentation of continental rather than marine character. This interpretation thus contrasts with the concept of an initial sag or ancestral Red Sea basin (Hutchinson and Engels, 1972; Coleman, 1972, personal communication); pre-drift sediments in the Red Sea area would have been of broad epicontinental sea origin (Lowell and Genik, 1972).

Though we doubt that sufficient information is available to satisfactorily resolve the question of arch versus sag, and any incipient rift system undoubtedly has elements of both, our reasons for preferring the former are:

1. We are impressed with the outcrop distribution of Precambian crystalline rocks which are extensively exposed around the perimeter of the Red Sea and define a foundered arch of large regional dimensions,

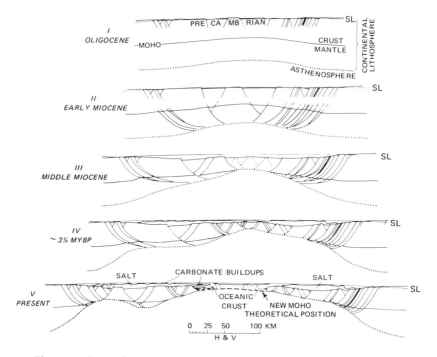

Figure 5. Sequential structural evolution of the southern Red Sea. Note that significant vertical movement and associated dip-slip are natural consequences of arching, breakup, and separation of a lithospheric plate and occur at all stages of the model. Published with permission of the American Association of Petroleum Geologists.

2,000 km long by 500 km wide on the average, with younger sedimentary strata dipping gently off the flanks (Fig. 6).

2. Normal faults bounding the Red Sea are parallel with the coastlines rather than perpendicular to the direction of separation (Fig. 7). The faults are interpreted to have received this trend very early, i.e., during an arching stage, wherein faulting was longitudinal to a broad regional arch (Cloos, 1939).

3. In areas adjacent to the southern Red Sea sedimentary rocks interbedded with basalts radiogenically dated as Oligocene (the postulated time of arching) are mainly of continental origin. This holds for the Dogali Series in Ethiopia, the Trap Series in Yemen (U.S. Geol. Survey, 1963; Geukens, 1966) and various units in Saudi Arabia (Karpoff, 1957; Brown, 1970). We reason that, had the Red Sea evolved initially as a sag, a marine incursion would have quickly fol-

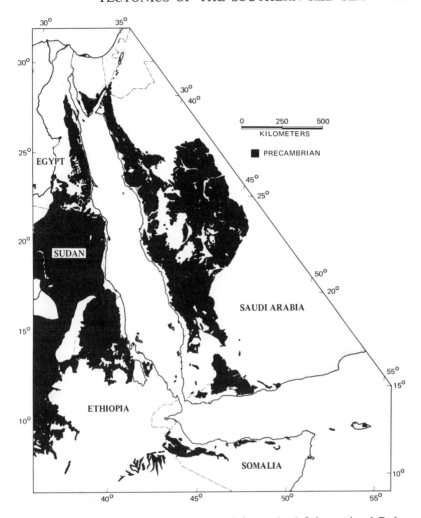

Figure 6. Outcrop distribution of Precambrian rocks defining regional Red Sea arch. Published with permission of the American Association of Petroleum Geologists.

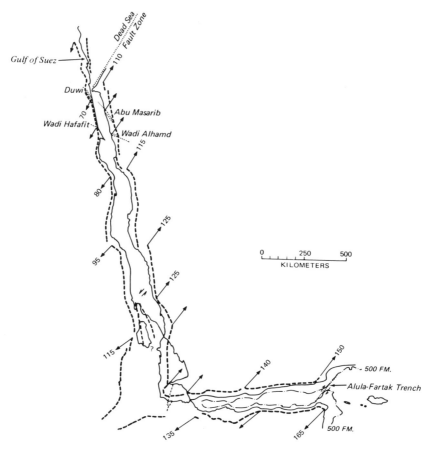

Figure 7. Pre-rift restoration of the Red Sea and Gulf of Aden. Heavy dashed line defines approximate limits of rifting. *Relative* directions and amounts of separation shown by arrows with accompanying values in kilometers. Published with the permission of the American Association of Petroleum Geologists.

lowed and marine sediments would be among the oldest preserved owing to early deposition and burial.

Results in the B1 well (Fig. 3) are interesting in this regard. The well encountered basalt at 1323 m (K/Ar whole rock date 4.9 ± 0.5 MY, $K_2O\%$ 1.15) and drilled interbedded salt and basalt flows to total depth of 2965 m (K/Ar average date 30 MY, $K_2O\%$ 0.206), the latter predominating below 2018 m. The basalts were intensively altered, particularly in the lower part of the hole where leaching during metasomatism caused significant potassium loss. Most of the volcanic fragments in the ditch cuttings are of olivine basalt variety; no fresh olivine was found but its original presence is indicated by serpentine pseudomorphs. Iron in the form of magnetite and ilmenite is locally abundant.

We postulate from the foregoing information that Oligocene volcanism was of alkalic continental affinities and that oceanic tholeiites are not present at the well location but only in the much younger axial trough to the east. The salt interbeds of Oligocene age may have accumulated in a continental sebkha or lake environment which became marine in the subsequent rifting and subsidence stage in Miocene time. We suspect that the B1 location is underlain by greatly thinned continental crust. The initial arching stage of the Red Sea seems to us analogous to the present uplifted condition of the Ethiopian, Kenyan, and Tanzanian parts of the East African rift system, in that the latter have lakes in their fault-bounded valleys and substantial volcanism in a setting which is distinctly continental, rather than oceanic.

The difficulty in determining an initial arch versus sag origin for the Red Sea is also compounded by the fact that once initiated an arch almost immediately begins to subside in its crestal portion by displacement on downward propagating normal faults. Grabens and fault troughs thus appear very early in the history of arching as sites of continental sedimentation. An arch of perfect geometric form is never achieved, rather, the central portion subsides.

As a result of early crestal faulting, continental sandstones of reservoir potential may have formed when short streams drained developing normal fault blocks and scarps. The concept of potential reservoir-quality sandstones being present in non-marine clastic wedges formed during arching seems to be fulfilled by sandstones in the Dogali Series of Ethiopia, the Trap Series of Yemen, and the Shumazi Group of Saudi Arabia. Potential reservoir rocks of pre-arch origin might also be present. Porous sandstones were deposited in a Jurassic epicontinental sea that covered much of Ethiopia, Yemen, and Saudi Arabia (Gillmann, Letullier, and Renouard, 1966).

Arching is presumed to have been caused by a thermally activated volume increase attendant to dynamic convective upwelling in the asthenosphere.

RIFTING

After the arching stage, divergent convective flow in the asthenosphere is interpreted to have thinned the continental lithosphere, in large part by shear on normal faults, which in this general stress regime would propagate upward through the brittle part of the lithosphere.[5] This heralded the rifting stage in early and middle Miocene time (Fig. 5, sections II and III). It was marked by pronounced subsidence of horsts and grabens, an extensive marine incursion possibly inaugurated by a transgressive basal sandstone, and the accumulation of a thick evaporite sequence.

Insight into the environment of the Red Sea Miocene evaporite accumulation may be provided by the Danakil Depression, where at least 1 km of halite of marine origin and Quaternary age has been drilled in the deep asymmetric west side of the basin (Fig. 8; Holwerda and Hutchinson, 1968). The Danakil Quaternary salt accumulated in a narrow, restricted rifted trough in a hot, arid, low-latitude setting just as did the Red Sea Miocene salt. Both salts must have been precipitated by evaporation of sea water, which was externally replenished from the Red Sea through the Gulf of Zula into the Danakil Depression during the Quaternary and from the Mediterranean through the Gulf of Suez into the Red Sea during the Miocene. The stratigraphic relationship of the Danakil Quaternary salt above the continental clastics of the Mio-Pliocene Desset series (Lowell and Genik, 1972) is identical with that of the Miocene Red Sea salt overlying older continental clastics. Unless the Danakil Depression was at much greater depths in the recent past than it is at present, at least the uppermost part—if not the whole of the salt sequence—must be of very shallow water origin. Because of the close similarity of the Danakil evaporite occurrence to that of the Red Sea, we are tempted to assign a comparable shallow-water origin to the latter. In any case, both salt basins formed as a consequence of continental breakup, and they must have many counterparts in the geologic record.

DOUBLE RIFTING

The inception of a second rift west of the axial-trough rift began in the Pliocene (Fig. 5, section IV). At the same time the upper part of

---

[5] Upward propagating listric normal faults play a more fundamental role in lithospheric thinning than do the earlier downward propagating faults (Fig. 5), but they need not reach the surface. Indeed in areas having monoclinal flexures more or less in the absence of faults, as along the coastal area of southwest Saudi Arabia (Coleman and Brown, 1971), the flexures may be controlled by upward propagating faults at depth.

Figure 8. View west of escarpment of Ethiopian Plateau from above Dallol. Escarpment with rotated normal fault blocks in background. Salt in middle ground, while below wing of aircraft at Dallol some of salt has been uplifted by an igneous feature from below.

the continental crust in the axial trough was almost sufficiently thinned to allow insertion of oceanic basalts that have magnetic anomalies dated to 3.5 MY, or late Pliocene (Vine, 1966). Sea floor spreading continues at present (Fig. 5, section V), and two rifts are active.

Volcanism has accompanied each stage of the structural evolution of the southern Red Sea area and is active at present.

AMOUNT OF SEPARATION

The structural evolution summarized in this paper has an important bearing on the amount of separation that has taken place in the Red Sea. The amount of separation in turn, has a fundamental implication for petroleum occurrence.

Our studies have shown that the breakup of continental lithosphere was by normal faulting, which dictates that sialic blocks must be present seaward of the present coastlines (Fig. 9); thus, the basement from coast to coast cannot be oceanic crust. The significance of this observation to petroleum exploration is that hydrocarbons are much more likely to occur in areas underlain by continental crust than in areas underlain

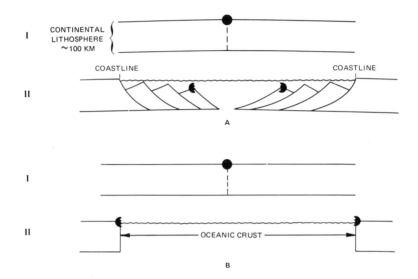

Figure 9. Diagrammatic sections illustrating how continental breakup by normal faulting does not permit pre-breakup fit for Red Sea at present coastlines. *A*. During breakup of lithosphere by normal faulting, present coastlines were never contiguous. Original point of contact is now submerged. *B*. Only if vertical fault surface is developed and maintained can present coastlines once have been contiguous. Published with the permission of the American Association of Petroleum Geologists.

by oceanic crust. The argument against oceanic crust flooring the Red Sea from coast to coast is further substantiated by magnetic and seismic reflection and refraction data. First, the central part of the Red Sea shows large-amplitude magnetic anomalies commonly associated with oceanic crust, whereas on the west the smooth magnetic profile suggests that this area is underlain by continental crust (Fig. 3). The difference in magnetic character cannot be attributed to a change in sediment thickness, because the latter remains almost constant in the vicinity of the change in the magnetic profile. Second, the basement west of the western rift on reflection seismic sections (Fig. 4) is almost certainly continuous with Precambrian crystalline basement farther west on shore. The top of the basement can be correlated on either side of the western rift, which strongly suggests that Precambrian basement is also present east of the rift. Finally, numerous refraction lines in the Red Sea have continental crustal velocities (Girdler, 1969), indicating that blocks of this composition are present seaward of the coastlines.

The present coastlines of the Red Sea thus cannot be used to make a pre-rift restoration, and neither can a fit be made with isobaths. The northern part of the Red Sea seems to be floored continuously by continental crust; consequently, there can be no isobath that marks the boundary between continental and oceanic crust. In the southern Red Sea where such a boundary is present, it is masked by shoaling reefs and salt structures. Consequently, no single pair of bathymetric contours defines the breaks between continental and oceanic crust in the Red Sea as they do on more mature continental margins including the Gulf of Aden. To make a pre-rift restoration of the Red Sea, we have therefore defined the limits of rifting between which essentially all the normal faulting, thinning, and separation have taken place, and beyond which the continental lithosphere has moved as part of a relatively rigid plate (Fig. 7).

The restoration has been made by using 110 km of left-lateral strike slip along the trend of the Dead Sea fault zone (Freund et al., 1970, for summary), matching Precambrian shear zones (Duwi and Wadi Hafafit in Egypt with Abu Masarib and Wadi Alhamd, respectively, in Saudi Arabia) across the northern part of the Red Sea (Abdel-Gawad, 1969; Fig. 10) using approximately 70 km of movement *oblique* to the trend of the Suez graben for extension of that normal-faulted feature, and using for azimuth of movement the trend of the Dead Sea fault zone and earthquake focal plane solutions in the southern Red Sea (Isacks et al., 1968) and along the Alula-Fartak trench in the Gulf of Aden (Laughton, 1966).

Continental breakup and dispersal of the Red Sea have no doubt been accomplished by strike-slip or transform faulting in addition to rifting (normal faulting). The focal-plane solution in the southern Red Sea cited above shows lateral motion, as does another in the central Red

Figure 10. Gemini XII photograph (looking south) showing position of Precambrian shear zones (Abdel-Gawad, 1969). Note that the course of the Nile River on the west flank of the Red Sea arch is an example of drainage deflected by arching such that relatively little clastic deposition occurs in the rifted trough and evaporites are free to accumulate. Note also that the Suez graben, though on trend with the Red Sea and the site of considerable Miocene rifting and evaporite accumulation (Heybroek, 1965), may be an aborted rift, for it is not now on the plate boundary associated with the Red Sea; that boundary turns up the Gulf of Aqaba into the Dead Sea fault zone instead of following from the northern Red Sea into the Gulf of Suez. The Suez graben is, however, in a very favorable tectonic setting for oil occurrence. It is a Miocene cratonic basin in which subsidence has led to large normal fault blocks, some with reservoir rocks present under an evaporite seal, that contain major oil fields. High temperature has been neither a geologic nor a drilling factor in Suez graben petroleum development. This is attributed to the cessation of extension (which ultimately leads to active sea-floor spreading) when the lithosphere was still of almost normal thickness (cf. Fig. 13). Black and white reproduction of Gemini photo No. S-66-63481, courtesy of NASA.

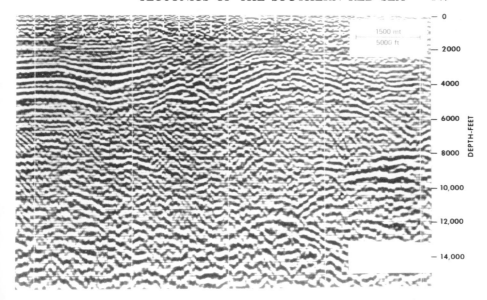

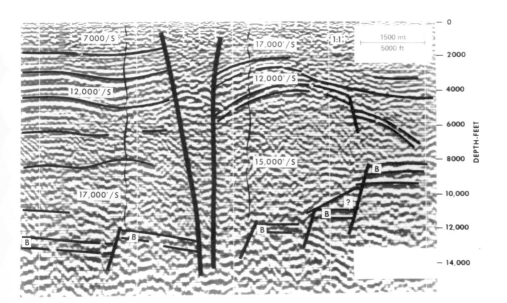

Figure 11. Reflection seismic section from former Esso-Mobil concession area in Ethiopian waters of the southern Red Sea showing fault interpreted as having had strike-slip on the basis of mismatched velocities and seismic reflections. (B-basement)

Sea (Fairhead and Girdler, 1970). Allan (1970) and Phillips (1970) on the basis of offset magnetic anomalies have proposed transform faulting in the central Red Sea centered at 21°N. In the former Esso-Mobil concession area of the southern Red Sea, we have interpreted a strike-slip fault trending northeastward in an orientation consistent with transform faulting (Fig. 11). Transform faults in the Red Sea are difficult to discern because the oceanic crust, on which they are usually best developed, ranges from relatively narrow in width in the south to absent in the north. As the Red Sea continues to evolve into an ocean, transform faults should develop accordingly and be more easily discerned, as they are in the Gulf of Aden, where oceanic crust is wider (Fig. 2).

Controls on Petroleum Occurence

We have shown that the dominant structural styles—and those most likely to form traps—in the southern Red Sea are normal faults and salt features, both of which are directly related to breakup of lithospheric plates. The intimate association of normal fault and salt structural styles is illustrated by the reflection seismic section of Figure 12, which shows flowage in a piercement salt feature that may have been triggered by normal faulting below.

In addition to domes, walls of salt can be traced for tens of kilometers on either side of the southern part of the Red Sea (Figs. 3, 4; Gillmann, 1968). The salt walls are invariably downdip from areas of salt withdrawal, above which downbending of sediments has occurred. All of these salt structures, domes, walls, and downbends are capable of forming structural reversals. Unfortunately, plays in targets above the salt in the Red Sea have not been encouraging, probably because source and seal rocks do not seem to be very plentiful.

Rather than providing structures, the salt might be more important in being ideally situated to seal hydrocarbons in underlying porous reservoirs. Normal fault blocks can be expected to form traps having porous reservoirs sealed by salt above (Fig. 3). Reservoirs could be of three origins: pre-rift; continental clastics (which might be of somewhat limited lateral extent) associated with arching; and basal sands associated with the marine transgression of the rifting stage. The subsalt play, however, has been significantly affected by the double-rifting mode of continental lithospheric breakup.

Double rifting causes extension of continental crust over a broader

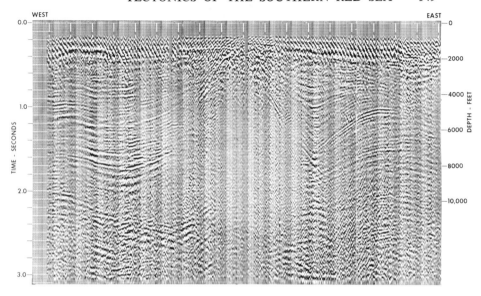

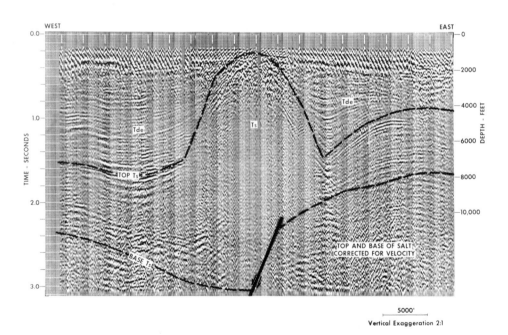

Figure 12. Reflection seismic section from former Esso-Mobil concession area in Ethiopian waters of the southern Red Sea showing piercement salt dome with normal fault below. (See Fig. 3 for stratigraphic legend.)

area than does single rifting, which effects a narrower tapered crustal wedge (Fig. 13). Assuming a deep heat source, double rifting with its

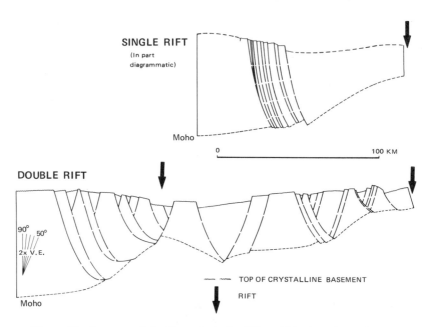

Figure 13. Contrast of double and single rifting. Thinning of crust occurs over a wider area by double rifting than by single rifting and apparently results in higher heat flow from sublithospheric sources for the former. Note that one type of origin for a "shelf-edge high" could be by double rifting, wherein a high at the continental margin is left between two rifts. (The double-rifted section is the west side of the structure section of Figure 3.)

thinner continental crust would lead to higher temperatures in the sedimentary section. Heat flow in the southern Red Sea apparently increases toward each rift from a cooler intervening area (Fig. 14). Moreover, heat flow is very high, often greater than 4.0 microcalories, and geothermal gradients in the southern Red Sea range from 28° to more than 50° C/km (2.5°–3.7°F/100 ft). In the thermal maturation process, oil subjected to this kind of heat might soon be converted to gas. A gas blowout beneath the salt in the C1 well partially confirms that the area may be gas prone, at least at depths of 3 km and greater. On the other hand, oil seeps have been reported from the southern Red Sea presum-

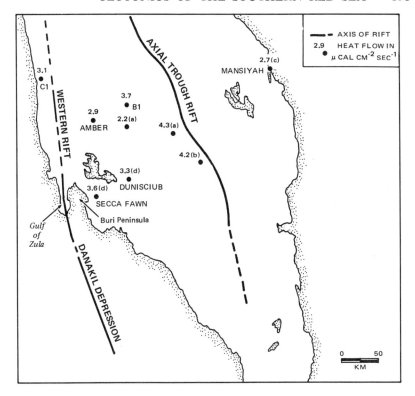

Figure 14. Heat flow in southern Red Sea. Values are (a) from Langseth and Taylor, 1967, (b) from Sclater, 1966, (c) from Girdler, 1970, and (d) calculated from Frazier, 1970.

ably emanating from sediments at relatively shallow depths above the Miocene salt. In this case high temperatures may have hastened the maturation process of organic material to oil (Klemme, this vol.).

CONCLUSION

The southern Red Sea demonstrates a close relationship between plate tectonics and the controls on occurrence of petroleum. We believe that many of these relationships can no doubt be applied to other lithospheric plates which, if geothermally cooler, might be more prospective for hydrocarbons and at the same time easier to drill.

REFERENCES

Abdel-Gawad, M., 1969. New evidence of transcurrent movements in Red Sea area and petroleum implications, *Am. Assoc. Petroleum Geologists Bull.*, v. 53, no. 7, pp. 1466–1479.

Allan, T. D., 1970. Magnetic and gravity fields over the Red Sea, *Royal Soc. London Philos. Trans.*, ser. A., v. 267, no. 1181, pp. 153–180.

Brown, G. F., 1970. Eastern margin of the Red Sea and the coastal structures in Saudi Arabia, *Royal Soc. London Philos. Trans.*, ser. A, v. 267, no. 1181, pp. 75–87.

Coleman, R. G., and Brown, G. F., 1971. Volcanism in Southwest Saudi Arabia, *Geol. Soc. America*, Abs. with Programs, v. 3, no. 7, p. 529.

Cloos, H., 1939. Hebung-Spaltung-Vulkanismus: Elemente einer geometrischen Analyse irdischer Grossformen, *Geol. Rundschau*, Bd. 30, Heft 4A, pp. 405–527.

Fairhead, J. D., and R. W. Girdler, 1970. The seismicity of the Red Sea, Gulf of Aden, and Afar triangle, *Royal Soc. London Philos. Trans.*, ser. A, v. 267, no. 1181, pp. 49–71.

Frazier, S. B., 1970. Adjacent structures of Ethiopia—that portion of the Red Sea coast including Dahlak Kebir Island and the Gulf of Zula, *Royal Soc. London Philos. Trans.*, ser. A., v. 267, no. 1181, pp. 131–141.

Freund, R., Z. Garfunkel, I. Zak, M. Goldberg, T. Weissbrod, and B. Derin, 1970. The shear along the Dead Sea rift, *Royal Soc. London Philos. Trans.*, ser. A, v. 267, no. 1181, pp. 107–130.

Geukens, F., 1963. Geology of the Arabian Peninsula Yemen, *U.S. Geol. Survey Prof. Paper*, no. 560-B, 23 pp.

Gillmann, M., 1968. Primary results of a geological and geophysical reconnaissance of the Jizan coastal plain in Saudi Arabia, *Second Regional Technical Symposium*, Soc. Petroleum Engineers A.I.M.E., Dhahran, Preprint, 25 pp.

Gillmann, M., A. Letullier, and G. Renouard, 1966. La Mer Rouge géologie et probleme petrolier, *Inst. Français Petrole Rev.*, v. 21, no. 10, pp. 1467–1487.

Girdler, R. W., 1969. The Red Sea—A geophysical background. In Degens, E. T., and Ross, D. A., eds., *Hot Brines and Recent Heavy Metal Deposits in the Red Sea*, New York, Springer-Verlag New York Inc., pp. 38–58.

———, 1970. A review of Red Sea heat flow, *Royal Soc. London Philos. Trans.*, ser. A, v. 267, no. 1181, pp. 191–203.

Heirtzler, J. R., 1968. Sea-floor spreading, *Sci. American*, v. 219, no. 6, pp. 60–70.

Heybroek, F., 1965. The Red Sea Miocene evaporite basin. In *Salt Basins around Africa*, London, The Institute of Petroleum, pp. 17–40.

Holwerda, J. G., and R. W. Hutchinson, 1968. Potash-bearing evaporites in the Danakil area, Ethiopia, *Econ. Geology*, v. 63, pp. 124–150.

Hutchinson, R. W., and Engels, G. G., 1972. Tectonic evolution in the southern Red Sea and its possible significance to older rifted continental margins: *Geol. Soc. America Bull.,* v. 83, pp. 2989–3002.

Isacks, B. L., J. Oliver, and L. R. Sykes, 1968. Seismology and the new global tectonics, *Jour. Geophys. Research,* v. 73, pp. 5855–5899.

Karpoff, R., 1957. Sur l'existence du maestrichtien au nord de Djeddah (Arabie séoudite): *Acad. Sci. Comptes Rendus,* v. 245, no. 16, pp. 1322–1324.

Klemme, H. D., Geothermal gradients, heat flow and hydrocarbon recovery (this volume).

Langseth, M. G., and P. T. Taylor, 1967. Recent heat flow measurements in the Indian Ocean, *Jour. Geophys. Research,* v. 71, pp. 5321–5355.

Laughton, A. S., 1966. The Gulf of Aden, *Royal Soc. London Philos. Trans.,* ser. A, v. 259, no. 1099, pp. 150–171.

————, 1970. A new bathymetric chart of the Red Sea, *Royal Soc. London Philos. Trans.,* v. 267A, no. 1181, pp. 21, 22.

Laughton, A. S., R. B. Whitmarsh, and M. T. Jones, 1970. The evolution of the Gulf of Aden, *Royal Soc. London Philos. Trans.,* ser. A, v. 267, no. 1181, pp. 227–266.

Lowell, J. D., and G. J. Genik, 1972, Sea-floor spreading and structural evolution of southern Red Sea, *Am. Assoc. Petroleum Geologists Bull.,* v. 56, no. 2, pp. 247–259.

Phillips, J. D., 1970. Magnetic anomalies in the Red Sea, *Royal Soc. London Philos. Trans.,* ser. A, v. 267, no. 1181, pp. 205–217.

Sclater, J. G., 1966. Heat flow in the northwest Indian Ocean and Red Sea, *Royal Soc. London Philos. Trans.,* ser. A, v. 259, no. 1099, pp. 271–278.

U.S. Geol. Survey, Arabian American Oil Co., 1963. Geology of the Arabian peninsula: Map I-270A Kingdom of Saudi Arabia, Ministry of Petroleum and Mineral Resources, and U.S. Department of State.

Vine, F. J., 1966. Spreading of the ocean floor: new evidence, *Science,* v. 154, no. 3755, pp. 1405–1415.

Having dwelt extensively on the subject of basins, we now pass to the processes that fill them with sediment, and to the nature of this fill. Joseph Curray, one of the leaders in the study of recent marine sediments, provides us with an overall perspective of marine sedimentation, stresses the problems involved in applying sedimentary patterns of the Quaternary epoch to the geological past, and offers a geosynclinal classification based on present sedimentation patterns related to plate tectonics.

Marine Sediments, Geosynclines and Orogeny

Joseph R. Curray[1]

ABSTRACT

Terrigenous sediments entering the ocean encounter a series of sediment traps or depositional environments in the shore zone, continental shelf, continental slope and associated environments, and base-of-slope environments. Some sediments are temporarily or permanently deposited in these traps, while others may pass through them or bypass them to reach the ocean floor. Biogenic, authigenic, and volcanic constituents may be added at any stage along the way and may dominate the character of the sediment mixtures.

Many surficial sediments in the oceans, and especially on continental shelves, are relict from the lowered sea levels of the Pleistocene and are not in equilibrium with present environmental conditions. Some are being modified today in their present environments and have attained a state of pseudo-equilibrium, but they are not being nor could they be transported from their sources nor deposited today. Similarly, morphology of the shelf, slope, submarine canyons, base-of-slope environ-

[1] Scripps Institution of Oceanography, University of California at San Diego, La Jolla, California.

ments, and slope stabilities are geologically abnormal today as a result of the Pleistocene. In order to apply uniformitarianism and interpret ancient sediments and environments in terms of modern oceans, relict and modified relict sediments and morphology must be delineated, interpreted and eliminated from comparison. We should compare ancient sediments and environments only with those in the modern ocean where all scars of the Pleistocene have been healed.

Environments and their depositional facies are arranged in different combinations depending upon meteorological and oceanographic conditions, continental margin and ocean floor morphology, and plate tectonic setting. With sufficient time and sediment accumulation, some assemblages of environments may become actualistic equivalents to geosynclines of the past. Formation of such "geosynclines" and accumulation of geosynclinal volumes of sediment are not related to their subsequent orogenic fates by a predestined geotectonic cycle. With the exceptions of continental margin and island arc subduction zone "geosynclines," orogeny is caused by subsequent random events of plate tectonics. Eight different actualistic geosynclinal types and seventeen subsequent orogenic fates or tectonic changes are considered to be the major possibilities.

I am grateful to many colleagues for assistance in preparation of this manuscript: many of the ideas and concepts it contains have come from discussions, especially with Dr. David G. Moore. I have been supported in part by the Office of Naval Research, and have used space and facilities of the Geologisches Institut, Eidg. Technische Hochschule, Zürich, Switzerland, during preparation of the manuscript.

INTRODUCTION

This is perhaps an appropriate time to examine the relationship between the petroleum industry and the marine realm. Geologists affiliated with industry have made or inspired many of the advances of the geological revolution of the past decade, but the industry is operationally just now preparing to take its second giant step into the ocean. Cumulatively, the petroleum industry must have greater "experience," as measured by man-years, than the combined oceanographic institutions of the world and probably many navies, but, until now, that experience has been largely restricted to shore zone and inner continental shelf environments. The new concepts of tectonics, the technological advances of both exploratory and the JOIDES Deep Sea Drilling Project, and the increasing need for petroleum reserves make this second step off into deeper water imminent. Furthermore, our present realization, in contradiction with geological dogma, that deep water deposits are in fact common in and under both modern and ancient continental margins makes an understanding of these sediments and their facies relationships essential for exploration.

A voluminous literature already exists on the sediments of the oceans. An especially valuable compilation of information is contained in the reports of the JOIDES Deep Sea Drilling Project, particularly on pelagic and distal turbidite sediments. Many samples of the surface and near-surface sediments and rocks of continental margins have been described in the literature, and an even greater number of such detailed descriptions and interpretations of continental margin rocks exists in the files of petroleum companies operating offshore. In this short review, therefore, I will not attempt a comprehensive summary of the characteristics of marine sediments in general. I will instead deal only with certain aspects of modern environments of deposition and their sediments, peculiarities in these environments resulting from atypical conditions of the Quaternary, the relation to plate tectonics, and finally some possible paths of tectonic evolution of the major assemblages of these sedimentary facies.

MORPHOLOGY OF THE OCEANS AND PLATE TECTONICS

A first order subdivision of the crust of the earth into continent and ocean basin can be made geophysically. Another first order subdivision can be made by position on lithospheric plates or plate edges. Thus in Table I we have all possible environments of deposition of sediments of any kind, ensialic or ensimatic, orogenic or epeirogenic, marine or continental. Marine sediments, the concern of this review, can be deposited in any of these pigeon-holes, but we will be primarily interested in those of the continental margin and ocean floor. The classification of continental margin types and oceanic environments can be related both to this table and to the physiographic diagrams used to illustrate them. I am using the term *continental margin* in a broad sense to include everything on the edge of the continental crust affected by the proximity of the ocean basin, and everything on the edge of the oceanic crust

TABLE I

MAJOR TECTONIC SUBDIVISIONS OF THE EARTH

	CONTINENTAL CRUST			OCEANIC CRUST
MIDPLATE POSITION	CRATONIC	MIDPLATE CONTINENTAL MARGIN		OCEAN FLOOR
PLATE EDGE POSITION RIFT ZONE	RIFT VALLEY	NASCENT CONTINENTAL MARGIN		SPREADING RIDGE
SUBDUCTION ZONE	CONTINENTAL COLLISION MOUNTAIN RANGE	SUBDUCTION ZONE CONTINENTAL MARGIN		ISLAND ARC AND TRENCH COMPLEX
TRANSFORM FAULT ZONE	TRANSFORM FAULT CUTTING ACROSS CONTINENT	TRANSFORM FAULT TYPE CONTINENTAL MARGIN		ACTIVE PART OF FRACTURE ZONE

affected by the immediate proximity of the continent. This includes everything from a volcanic arc to the distal end of a continental rise.

Morphological Continental Margin Types

1. *Andean type,* or marginal depression type (Fig. 1), is typical of most of the Pacific, or almost half of the world's continental margins. Important sedimentary environments may include the trench, basins in the landward slope caused by irregularities in the slope, a continental shelf, sometimes an outer sedimentary ridge, a basin in the arc-trench gap (Dickinson, 1971a), the volcanic arc, and sometimes basins in the region behind the volcanic arc. The edge of the continental crust may lie under the continental slope, or it may lie well to the rear of the arc-trench gap. Purely oceanic island arcs contain the same provinces (Fig. 1).

2. *Atlantic type,* or the *shelf-rise* type of continental margin (Fig. 2), occupies a midplate position. It is typical of the Atlantic and parts of the Indian and Arctic Oceans, and when well developed to a "mature"

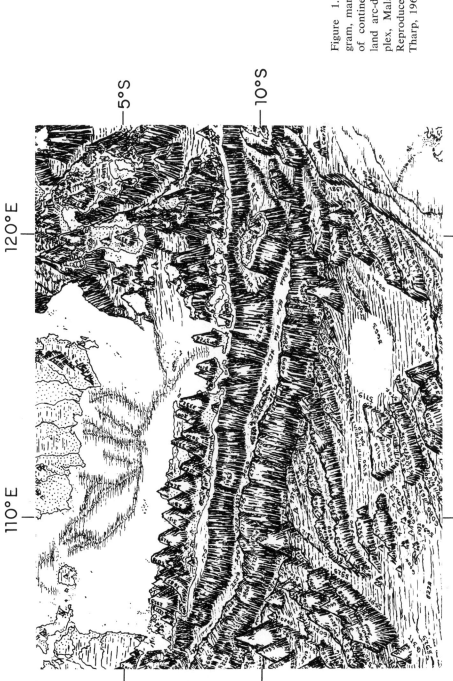

Figure 1. Physiographic diagram, marginal depression type of continental margin and island arc-deep sea trench complex, Malaysia and Indonesia. Reproduced from Heezen and Tharp, 1965, with permission.

stage by accumulation of sediment, is characterized by a wide coastal plain, continental shelf, and continental rise. Approximately half of the world's continental margins are of this morphologic type.

3. *The marginal plateau type,* best known by the Blake Plateau, (Fig. 2) is a subsidiary type characteristic of less than 5% of the world's continental margins. The plateau itself lies at the foot of a short plateau slope and is generally of intermediate depth, too deep to be considered part of the continental shelf (>550 m, Shepard, 1963a). The true continental slope or edge of the continental crust lies seaward of the plateau. Marginal plateaus appear to be characteristic of two different tectonic settings. The Blake Plateau, and others along the South American and Australian continental margins especially, lie in midplate positions, and are subsided fragments of the margin of the continental crust, commonly kept swept clean of sediments by oceanic currents. The second type, best described from northern California (Fig. 19) but common along other subduction-type continental margins, is formed by the scraping off of sediments from oceanic crust being subducted under continental crust. Slope basins thus formed are then filled with ponded sediments to produce plateaus. Conditions necessary for formation of such plateaus are relatively high ratio of rate of sediment supply to rate of subduction and an inability of the subduction zone to consume the sediment as well as the underlying oceanic crust.

4. *The continental borderland type* is best known for the southern California continental borderland (Fig. 3) but also appears to exist along the north coast of Venezuela and some other places. It differs from a marginal plateau in that its surface is broken into basins and interbasin ridges, some of which are islands. These two best examples appear to be related to transform fault zones, and might be considered as possibly characteristic of transform fault continental margins.

5. *Steep scarp type.* Some continental margins occuping midplate positions do not fit morphologically into other categories described. They display few of the characteristics of modification by accumulation of sediment, and would appear to be youthful or incipient Atlantic or shelf-rise types. Examples are the Iberian Peninsula and the Baja California side of the Gulf of California.

Oceanic Environments. Inasmuch as *continental margin* is used here in a broad sense to include wide segments of both the adjacent continent and ocean floor, the oceanic environments considered here are largely out of reach of terrigenous sediments, and are covered with pelagic sedi-

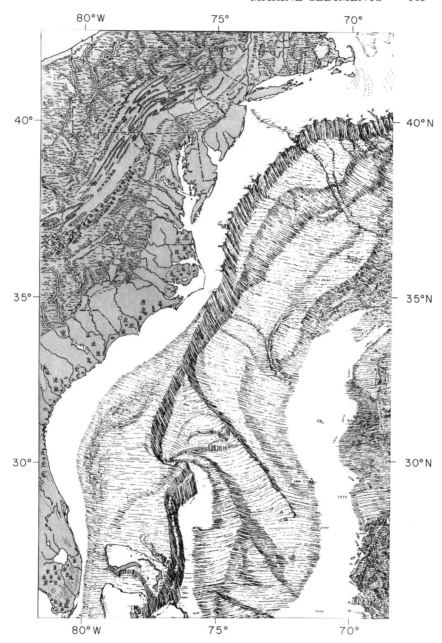

Figure 2. Physiographic diagram, east coast of the United States, showing shelf-rise and marginal plateau types of continental margins. Heezen and Tharp, 1968, with permission.

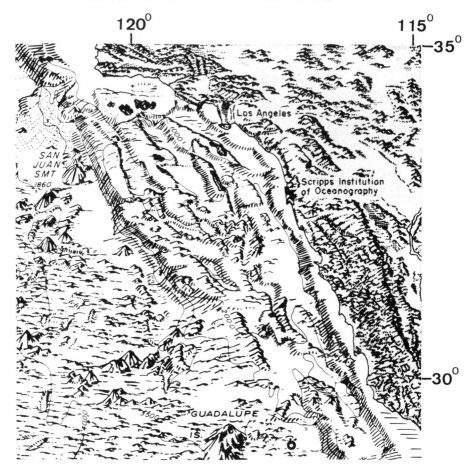

Figure 3. Physiographic diagram of southern California continental borderland, reproduced from Menard, 1964, with permission.

ments. The four oceanic plate positions shown in Table I will be considered in the review of sedimentary environments.

Uniformitarianism

We can easily describe modern oceans and sediments in terms of the morphological environments that exist today. It is not as easy to relate ancient sediments and environments to this present morphology, and we must be cautious in our attempts. Uniformitarianism, furthermore, does not imply a one-to-one correlation both forward and backward in geologic time. While not all ancient environments of deposition are represented in the modern world, it is probable that approximate equivalents to all modern environments have existed at some time during the geological past. Some of the environments we find today may, however, be geologically rare or almost unique to the Quaternary, because the Quaternary is certainly atypical of most of geological time. Insofar as possible, we should attempt to describe ancient environments and sediments in terms of modern counterparts, with as much elucidation as possible of the differences we infer. This translation is a two-step operation. We first must interpret modern sediments in terms of peculiarities of the Quaternary and what we should expect to have occurred in the analogous environments in pre-Quaternary time, or what will occur in the future in "post-Quaternary" time, barring another glaciation. Second, we must relate ancient sedimentary facies to their equivalent "non-Quaternary" environments and sediments.

QUATERNARY INFLUENCES

Quaternary influences include regional glacial and periglacial effects, climatic and circulation pattern changes, and eustatic fluctuations of sea level. Of these, we will concern ourselves primarily with the last. Oxygen isotope studies of deep sea cores (Emiliani, 1972) show apparent temperature fluctuations, which must be related to circulation pattern changes, with a periodicity of about 50,000 years for the past half million years. Changes in circulation affect paths of transport, and temperature changes affect productivity of the various biogenic constituents to pelagic oceanic sediments.

These climatic fluctuations have been accompanied by an indeterminate number of fluctuations of sea level (Fig. 4). It is obvious from the wide variation in published opinion that the chronology and magnitudes of the changes are controversial. Opinions differ because of different interpretations of the evidence, but also, and even more important, because of regional and local differences in stability. Only relative sea level changes can be observed. Distinction between eustatic and tectonic causes is a matter of interpretation and prejudice, and the many sources of error are difficult to avoid. It seems obvious to this writer and to many others who have worked on problems of late Quaternary sea levels that we are far from an understanding of eustatic sea levels. Real differences exist in relative stability of regions and whole continents that we had previously assumed were stable.

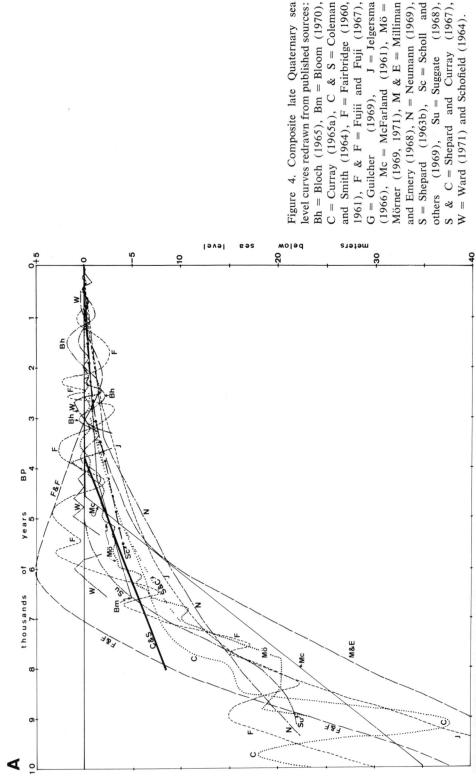

Figure 4. Composite late Quaternary sea level curves redrawn from published sources: Bh = Bloch (1965), Bm = Bloom (1970), C = Curray (1965a), C & S = Coleman and Smith (1964), F = Fairbridge (1960, 1961), F & F = Fujii and Fuji (1967), G = Guilcher (1969), J = Jelgersma (1966), Mc = McFarland (1961), Mö = Mörner (1969, 1971), M & E = Milliman and Emery (1968), N = Neumann (1969), S = Shepard (1963b), Sc = Scholl and others (1969), Su = Suggate (1968), S & C = Shepard and Curray (1967), W = Ward (1971) and Schofield (1964).

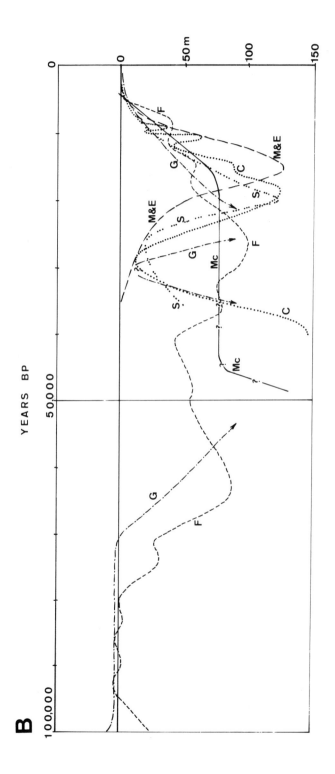

Two factors are much more important for our purposes here than an absolute chronology of sea level changes. First is the fact that there have been many large amplitude swings of sea level during the Quaternary. The second factor is that most published curves for sea level of the past 10,000 years show a pronounced decrease in the rate of rise at about 7000 BP. The effects of these influences are profound. Most shore zone environments were created only during the past 7000 years. Multiple transgressions and regressions have occurred back and forth across continental shelves during the Quaternary, affecting not only sediments and morphology of the shelves but also sediments and morphology of deeper parts of the continental margin and deep sea. During periods of lowered sea level or maximum regression, the shoreline, with its multiple sources of sediment, lay near the edge of the continental shelf, and greater amounts of sediment passed down the slopes into deeper water environments. These various effects will be discussed in the review of the individual environments of deposition.

MARINE SEDIMENTARY ENVIRONMENTS

Shore Zone

The shore zone (Fig. 5) is the ill-defined complex of environments of deposition of paralic or "marginal marine" sediments. The various individual environments we include here are beach, barrier, lagoon, estuary, tidal flat, and delta. Precise definitions are unimportant here and have been discussed elsewhere in somewhat more detail (Curray, 1969).

The shore zone environments of the modern oceans were created at various times within the past 7000 years, depending upon the slope of the surface across which transgression occurred, relative stability, intensity of oceanographic agents of sediment distribution, and rate of supply of sediments. Many transgressing shorelines, for example, stopped their landward advances and started building upward in place to form barriers or barrier islands (Fig. 6) after the decrease in the rate of rise of sea level (Fig. 4b). The locations and times of formation of barrier-lagoon systems appear to be related to three principal factors: first, they were formed during Holocene time only where rate of supply of sand to the shoreline was rapid. Where sand was particularly abundant, they formed 4000 to 7000 years ago, and the shoreline may have prograded subsequently. Where sand was somewhat less abundant, barriers formed more

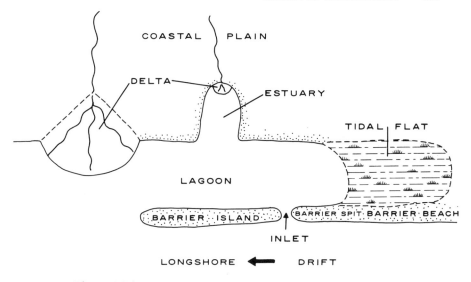

Figure 5. Diagrammatic representation of shore zone environments.

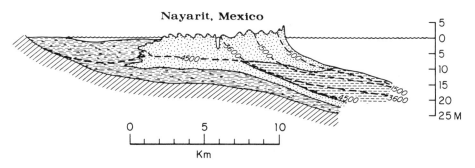

Figure 6. Section through the barrier-lagoon complex, Nayarit, Mexico, showing barrier prograding over the muddy facies of the inner continental shelf by formation of beach ridges. Barrier is underlain by lagoonal facies of the transgression and pre-transgressive alluvium. Time lines in years BP. Simplified from Curray, and others, 1969.

SUBAERIAL DELTA SHAPE

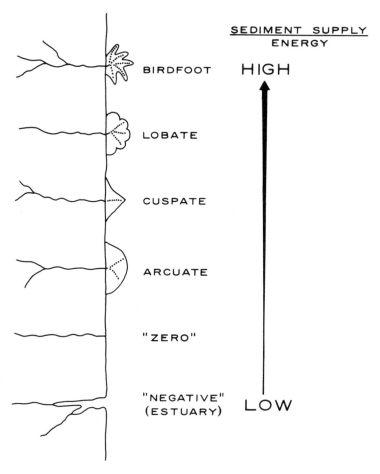

Figure 7. Subaerial delta morphology as a function of ratio between rate of sediment supply to energy in the environment for dispersal of the sediments.

recently than 4000 BP, and subsequent progradation is less likely to have occurred. Second, barrier-lagoon systems have formed mainly adjacent to wide, gently sloping continental shelves, where the rate of transgression was slower. And third, it appears that barrier-lagoon systems are predominantly located along those continental margins where the sea (Fig. 4b) has only recently attained its present relative level (Curray, 1969; Cromwell, 1971). These are the regions in the diagram with the lower sea level curves without pronounced fluctuations above and below present sea level.

Shore zone environments are very efficient sediment traps. Rivers entering the ocean drop much of their sediment load immediately with the decrease in current velocity, thus forming deltas. Subaerial delta form (Fig. 7) is in part a function of the balance between rate of sediment supply and efficiency of redistribution in the ocean (Bernard, 1965), and ranges from "bird-foot," lobate, cuspate, to arcuate, and then to what might be called the "zero" and "negative" deltas, or, in other words, "no delta" and "estuarine," respectively.

Estuaries (Fig. 8) generally trap more sediment derived from the open ocean and longshore drift from the seaward side than they release of river-derived sediment flowing into their landward terminations (Meade, 1969; Schubel, 1971). Little quantitative work has been done, but lagoons behind barriers probably do the same. Furthermore, tidal flats and the intertidal portions of estuaries and lagoons entrap sediments carried onto their surfaces by other well-described mechanisms (Postma, 1967), to further supplement the efficiency of the shore zone in trapping sediment as it tries to pass the interface from land to ocean.

The formation of modern shore zone environments has been recent, and they are continuing to evolve rapidly today. Projections suggest that many of the estuaries and lagoons of the world will be filled within the next few hundred to few thousand years. Many will evolve first to tidal flats, depending upon tidal range, and finally to subaerial alluvial plains. Deltas will start to form off the mouths of some of the rivers which will then enter directly into the sea. Existing deltas will continue to grow, forming and abandoning subdeltas, to prograde the coastline with broad deltaic complexes, until ultimately many rivers will debouch directly on the upper continental slope as they did during Pleistocene lowered sea level.

This future evolution, if we ignore the possibilities of a returning glacial stage or the influences of man, will represent a return to what we might term "normal" equilibrium conditions. The Holocene is atypical of most of geological time because of the recency of an extremely rapid transgression, which came directly on the heels of a rapid regression, etc. The shore zone today must certainly not be typical of most of geological time. We might ask how common were barrier-lagoon systems or estu-

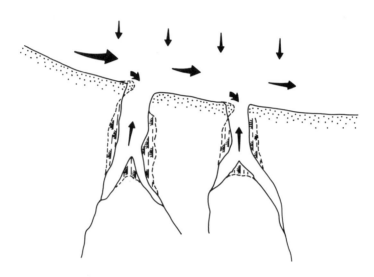

Figure 8. Diagrammatic estuary systems as sediment traps. Much of the river-derived sediment is retained, and some sediments from longshore drift and the open shelf are also drawn into the estuaries.

aries during most of geological time? They must have existed, judging from some of the excellent published studies in which such environments have been identified, but they must have been much less common than today, and were probably restricted only to certain tectonic, oceanographic, and sedimentological settings (subsiding continental margins?). Perhaps tidal flats were relatively more common than today, but they might also have been concentrated along subsiding continental margins. With longer periods of relative sea level stability, deltas must have been larger and more mature, and many must have prograded to the upper continental slope.

One of the important problems of the shore zone is the question of how much and what grain sizes of sediments are lost at the outer boundary to the continental shelf, and by what mechanisms. Fine sediments (mud, or silt and clay) certainly do escape from the shore zone, as bottomsets of deltas, and as lenses of mud related to each river source along the shoreline. Sands are more controversial. Several mechanisms have been proposed by which sands can escape from the nearshore, or surf-dominated zone along the shoreline, but a review of the evidence and the modern geologic record leads the writer to the strong belief that little sand leaks out, except by turbidity currents through the axes of submarine canyons. Under rather unusual circumstances, such as strong tidal currents parallel to shore and unusual storm wave action, some sand may escape, but in most circumstances, the net transport of sand is toward the shore zone, rather than away from it.

Continental Shelf

Continental shelves today are probably more atypical, as compared with most of geologic time, than any of the other environments to be discussed. The morphology of present shelves was most certainly strongly modified, if not largely shaped, by the repeated transgressions and regressions of the Pleistocene, although this idea has been recently disputed (Fairbridge, 1970). Surface sediment distribution is similarly out of equilibrium with existing environmental conditions.

The continental shelf is the submerged seaward-sloping surface of the edge of the continental mass, extending from the outer edge of the shore zone to an increase in gradient, which occurs at a worldwide depth average of about 130 m (Shepard, 1963a). This "shelf break" divides the continental shelf from the continental slope, or sometimes from a plateau or basin slope. Depth of the shelf break ranges widely from a few tens of meters to several hundred meters, with these deeper breaks coinciding rather well with the margins of glacially depressed continents. Even with this wide range, the variance around the mean is rather small today because of the control by the multiple fluctuations of sea level to approximately this depth. Shelves must have existed in pre-Quaternary time, but must not have been as uniform in depth, nor displayed as district shelf breaks as they do today. For this reason, we should apply the term *continental shelf* with some caution to ancient equivalents.

Surface sediment distribution on shelves (Fig. 9) is also out of equilibrium. Shepard (1932) was one of the first to realize that much of the surface of the shelf, and particularly the outer portions, are covered

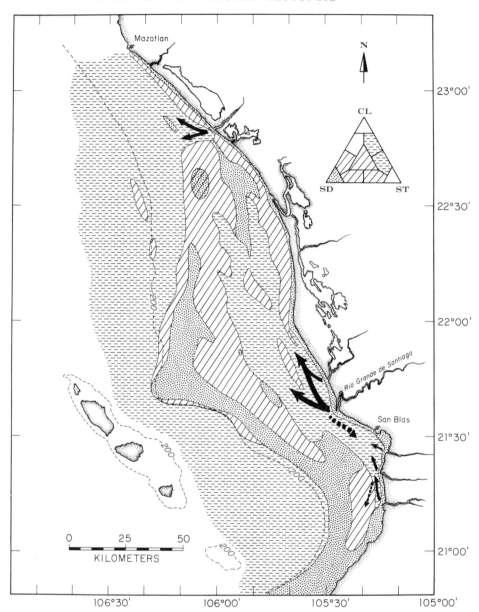

Figure 9. Surface sediment distribution, continental shelf off Nayarit and southern Sinaloa, Mexico. Outer shelf is covered with relict and palimpsest shore zone sediments and some pseudo-equilibrium sands in sand waves. Modern shelf facies muds are deposited only on the inner shelf.

with sandy sediments deposited in a shore zone environment during Pleistocene lowered sea level. Emery (1952) defined these as *relict* sediments, remnant from a different earlier environment, and he distinguished them from *detrital* sediments, often called *modern* or *equilibrium* sediments, and from *residual, authigenic,* and *organic* sediments. He later (1968) added another type, *volcanic.* A recent reconsideration by Swift and others (1971) has discussed some of the difficulties with this scheme in that many relict sediments have been reworked and are in fact adjusting to the environmental conditions in which they are found today, even though they were not transported to where they are under present environmental conditions. They added yet another term, *palimpsest,* to designate those reworked sediments that now have taken on characteristics of their present environment, but yet retain some of the characteristics of their original environment.

Such considerations might seem academic. They are in fact, of great geologic importance for the proper interpretation of ancient sediments. For example, we find sand blanketing the outer continental shelf today (Fig. 9). Can these sands be compared with what have been interpreted as shelf sands of ancient sediments? Sand waves, banks, and ridges in the strong tidal current regime on the shelf around Britain (Fig. 10) are compared with similar deposits of the past. I believe that seaward transport of sand beyond the shore zone is one of the very important problems in sedimentary geology today. Can sand leak out to the continental shelf, or are we misinterpreting the environments of many ancient sand bodies? In the writer's opinion, we have no known mechanism or process today that can account for sands in significant quantities on the continental shelf, other than by transgression with a low rate of supply of mud.

For this discussion, I will revise some of the terminology as used by Emery (1952, 1968), Curray (1965a), and Swift and others (1971). Rather than intermix terms describing the relationship of sediments to their environments with terms denoting sediment source, I will restrict my terminology to the former category.

Relict sediment is used as defined by Emery, as "remnant from a different earlier environment" (1952, p. 1105), with the further restriction that the sediments, or at least the major constituents, are not and could not be supplied today. *Relict* can apply to sediments of biogenic, authigenic or volcanic origin, as well as terrigenous.

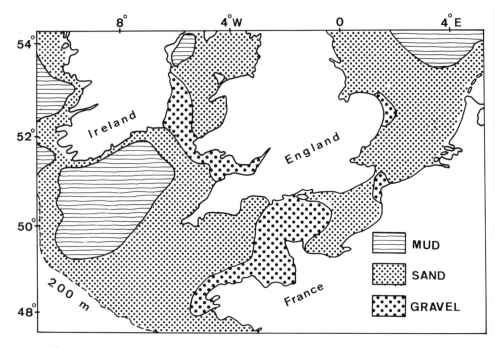

Figure 10. Predicted Holocene surface sediment distribution, continental shelf off Britain, as simplified from Stride (1963) and Belderson and others (1971). Much of the "gravel" area is basal conglomerate. Areas shown as "mud" have patches of mud apparently accumulating in depressions on a mixed sand and mud bottom. Much of the sand in the large areas shown as "sand" is in transport, either in sand ribbons in areas where tidal near-surface current velocities are two knots or more or as sand waves where velocities are lower. Much of this sand in transport will not accumulate where it is now. Sand is also in transport across some of the gravel or basal conglomerate areas. In my terminology, the muds are probably "equilibrium" sediments; some of the sands and gravels are "relict" or "palimpsest," and some are certainly "pseudo-equilibrium."

Equilibrium sediment is one that is in all respects related to its present environment, and is being or could be supplied and deposited today. It can be composed of terrigenous, biogenic, authigenic, or volcanic constituents, or any mixtures. There is no such thing as normal geologic time, but our objective here is to relate the sediments we find today to what we would have found in pre-Quaternary time, which must have been more nearly normal than the Holocene. Equilibrium sediments, then, are similar to what we might have found in these environments if the Quaternary had never occurred, or what might occur in the future after all scars of the Pleistocene have been healed.

Palimpsest sediment is a reworked or modified relict sediment that displays some characteristics of its present environment, while retaining some of the characteristics of its original environment (Swift and others, 1971).

Pseudo-equilibrium sediment is a relict sediment that is so extensively reworked that it has lost many of the characteristics of its original environment and exists in a state of equilibrium with its present environment in all respects except that it could not be formed or deposited today. Some authigenic, terrigenous, biogenic, or volcanic constituents may have been added, but the major constituents that dominate its character are relict, and are not being supplied or added today. In a sense, it is a relict sediment that is well adapted to and happy with its present environment but is not being formed in its present environment.

The relationships between these four sediment types are shown diagrammatically in Figure 11. Change from left to right can occur only by

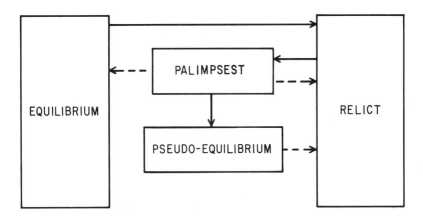

Figure 11. Relationship between shelf sediment types discussed in text.

change in the environment. Change from right to left occurs by reworking and/or by addition of constituents. Relict sediments can change to palimpsest, which can change either to pseudo-equilibrium by further reworking or to equilibrium by sufficient addition of constituents. Relict, palimpsest, and pseudo-equilibrium sediments may ultimately be buried by prograding blankets or wedges of equilibrium sediments.

Sediments of any marine environment may be classified in these terms, although they are most appropriate for sediments of the continental shelf. Most sediments of the shore zone are equilibrium sediments, although the environments are evolving rapidly. Emery (1968) estimates that 70% of the surface of continental shelves is covered with relict sediments. I would put the estimate a bit lower, and would subdivide the sediments into relict, palimpsest, and pseudo-equilibrium. Inasmuch as a continuum exists between these three types, it is impractical at this time to attempt a quantitative estimate. We might instead cite examples of each.

Rather pure unmodified relict sediments are well known on the shelves off both eastern United States and the Gulf Coast (Fig. 12), where faunal remains contained in the sands are almost entirely those of the original environments. These grade into more reworked regions where the palimpsest sands clearly contain faunal remains mixed from both original and present environments. Relict and perhaps locally palimpsest carbonate shelf sediments have been described from Florida (Fig. 13) (Gould and Stewart, 1955), and the Timor Sea (van Andel and Veevers, 1967). Finally, in the extreme are some of the sand waves of the shelf seas around Britain (Fig. 10) formed by unmixing and selective sorting from glacial deposits into clean sands containing only indigenous faunas.

Most of the surface sediments of continental shelves of the world are not yet in equilibrium from Pleistocene effects. In addition, the morphology is atypical. Unless the earth goes into another glacial stage, which it would most probably do barring either deliberate or accidental control by the activities of man, most of these shelves will ultimately return to equilibrium. Shelf facies muds will blanket the relict and reworked relict facies, although locally in regions of strong tidal currents this will be difficult and will require a long time. Sharp shelf breaks will be buried and eventually become shoaler and more rounded. Subsiding continental margins with strong currents and low rates of supply of sediments may not come into equilibrium, but may instead continue to subside as surfaces of nondeposition or low rates of deposition and become marginal plateaus.

If subsidence and shelf deposition keep pace with each other for a long time geologically, thick accumulations of shelf or "shelf basin" sediments can accumulate. A barrier (Fig. 14) helps to retain the sediments on the shelf and may be reef, volcanic, diapiric, or tectonic (Hedberg,

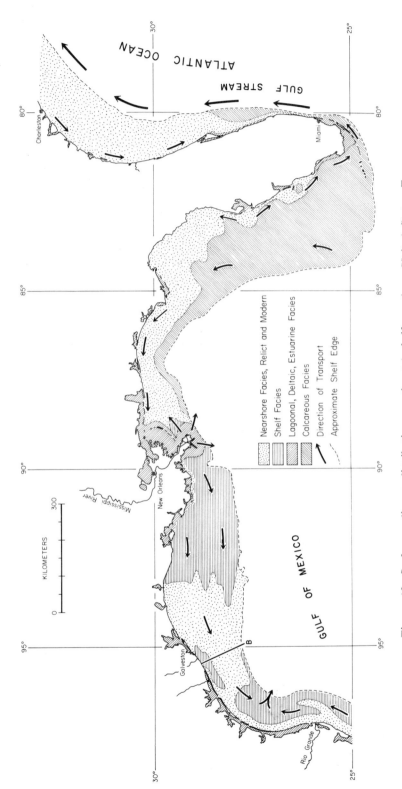

Figure 12. Surface sediment distribution, continental shelf, southern United States. From Curray, 1965a, as compiled from various sources.

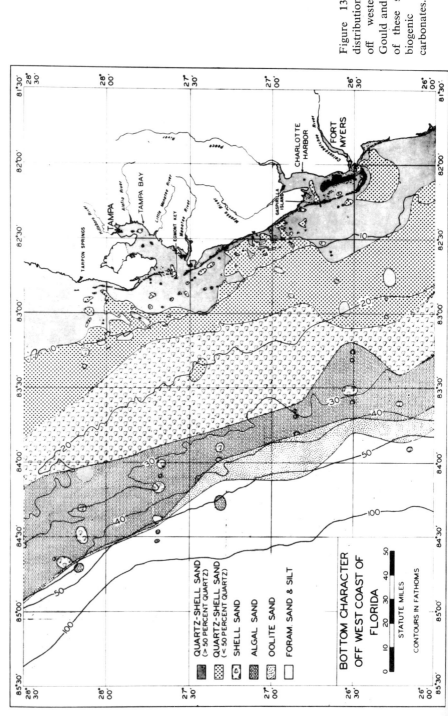

Figure 13. Surface sediment distribution, continental shelf off western Florida, from Gould and Stewart, 1955. Most of these sediments are relict biogenic and authigenic carbonates.

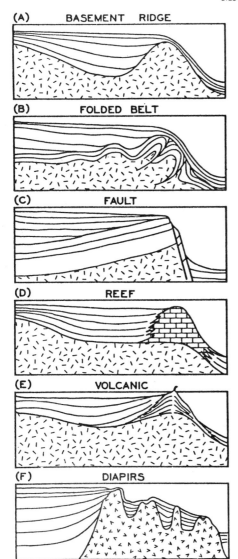

(A) BASEMENT RIDGE

(B) FOLDED BELT

(C) FAULT

(D) REEF

(E) VOLCANIC

(F) DIAPIRS

Figure 14. Diagrammatic representation of types of barriers at the outer margins of continental shelves, from Hedberg, 1970, with permission.

1964, 1970; Burk, 1968; Emery, 1969). The depositional surface may remain at shelf depths during subsidence and deposition, and thick sections of either terrigenous or carbonate miogeosynclinal sediments (*miogeoclinal,* Dietz and Holden, 1966) may be formed.

Continental Slope Environments

For convenience, the following environments will be grouped under this heading: continental slope, continental borderlands, marginal plateaus, slope basins, and submarine canyons. The various basin environments are sediment traps that are located on or near the continental slope proper or the edge of the continental crust. Their deposits range from insignificant to important thick sections.

The continental slope itself has an average gradient of only about 4°, and may or may not be the site of stable accumulation of sediments. Stability is a function of gradient, rate of accumulation, nature of the sediment, and subsequent disturbance (Moore, 1961). Rapid deposition off the deltas of large rivers may result in instability on slopes of less than 1°, while slow accumulation may result in stable deposits on much steeper gradients. Today the sediments that escape over the shelf break form a veneer of muddy sediments on the slopes of the world; but during Pleistocene lowered sea level, both sandy and muddy river sediments were deposited directly on the slope. Such rapid accumulation produced local instability and changed very many of the continental slopes of the world from depositional to erosional regimes. Repeated slumping in areas of high rate of supply, such as off river mouths or longshore drift regions of abundant supply, appears to have formed many of the submarine canyons of the world. Some canyons certainly did exist in pre-Quaternary time, just as they have existed generally throughout geologic time, but canyons are probably unusually abundant today. Canyons themselves may be filled, as recently reviewed by Stanley (1969), although their major importance is as conduits for supplying sediments to base-of-slope and ocean floor environments, as well as to the various sedimentary environments on the slope, such as plateaus, basins, and borderlands.

Structurally, slope environments such as basins and plateaus may be formed in many different ways, and are common on all different tectonic types of continental margins (in midplate, transform fault zone, or subduction zone positions). Morphologically, they may be interrelated simply by degree of filling and cover by sediments. A filled slope basin or borderland morphologically becomes a marginal plateau. And finally genetically, they may be related to shelves and shelf basins by subsidence not balanced by the same rate of deposition. Let us first consider the processes of sediment transport and deposition in these various possible environments, and then we will consider some modern examples, their structural origin, relation to plate position, and degree of modification by sedimentation.

Sediment fill or cover may be predominantly terrigenous, predominantly pelagic, or mixed. There are three important mechanisms of terrigenous sediment transport into these environments. Turbidity currents are perhaps the most important and are the only mechanism for providing significant amounts of sand or coarse sediments. They start to deposit their load at a decrease in gradient after descending the steeper slope or the axis of a submarine canyon. Rate and nature of basin fill is very directly related to supply by active open submarine canyons that head in or near shore zone sediment supplies. A fan and a fan valley system is normally formed at the foot of the canyon, and flat ponded basin fill occurs beyond the fan proper (Moore, 1969).

A second very important mechanism is turbid layer transport (Moore, 1969), as diffuse, low density, low velocity clouds of muddy water immediately above the sea floor (Fig. 15). Such turbid layers are known to be common on the continental shelf and are formed by resuspension of temporarily deposited fine sediments by long period waves, or by the activities of organisms. They move slowly along the bottom in response to both gravity and water currents, and must be extremely important in transporting large quantities of fine sediment across the outer shelf and to the slope. This is probably the most important single mechanism for the transport of sediments to continental slopes during times when the shoreline is not located along the shelf break. It is also important as a means of supplying fine sediments to the axes of submarine canyons traversing the shelf or slope, and finally as a means of direct supply of some fine sediments to nearshore basins and the landward margins of plateaus (Fig. 17).

The third mechanism for the transport of sediments on slopes is slumping and sliding. Slumped sediments have been recognized on continental slopes, in continental rises, and in various slope basin environments in many parts of the world (see, for example, Moore, 1961; Dott, 1963; Heezen and Drake, 1964; Curray, 1965b; Uchupi, 1967a; Emery et al., 1970; Moore et al., 1971). Cable breaks are attributed to such slumps, and the slumped sediments themselves have been compared with olistostromes and wildflysch (Fig. 16). Many slumps on present continental margins may be related to tectonic oversteepening or to oversteepening by rapid deposition during Pleistocene lowered sea levels. Perhaps slumping along midplate continental margins is more common today than normal.

Where rate of supply of terrigenous constituents is low, pelagic constituents predominate, usually either biogenic or authigenic. Lack of terrigenous supply may be related to climate, relief, and runoff from the adjacent continent, lack of a supply mechanism such as a submarine canyon system connected to terrigenous sources, topographic isolation, such as deeper water between the source and the environment, or prevention of terrigenous deposition by strong currents sweeping the sea floor. Important authigenic concentrations may result, such as phosphorites on the bank tops off southern California and many other places in the world, and manganese nodules on the Blake Plateau.

Slope basin and plateau environments may lie on any kind of continental margin tectonically. Midplate continental margin examples are the ponded basins formed by salt diapiric domes and ridges on the Gulf Coast (Moore and Curray, 1963; Uchupi and Emery, 1968; Lehner, 1969), or by shale diapirs off the delta of the Magdalena River, Colombia (Shepard et al., 1968). The Blake Plateau is in effect a filled basin behind a subsided barrier reef of Cretaceous age (see, for example, Uchupi, 1967b). As it subsided during early Tertiary opening of the Atlantic, it was kept swept clean of terrigenous sediment by the Gulf Stream. The Demerara Rise off Surinam is also a subsided shelf, although detrital sediments from the many large rivers of the adjacent coastline are now prograding across the inner plateau (Fig. 17).

Transform fault zone continental margins are best known in central and southern California (Emery, 1960; Moore, 1969) and northern Venezuela (Maloney, 1966). Continental borderlands were formed by transform fault emplacement of ridges of basement rock, with, in the case of the southern California continental borderland, accompanying block faulting. The basins created by the emplacement of basement rock ridges off central California are now filled (Fig. 18), and, in all of the area except the Arguello Plateau, prograded sediment has buried even the marginal plateaus thus formed (Curray, 1965b; Hoskins and Griffith, 1971; Silver et al., 1971).

The plateaus, basins, and arc-trench gap environments of subduction

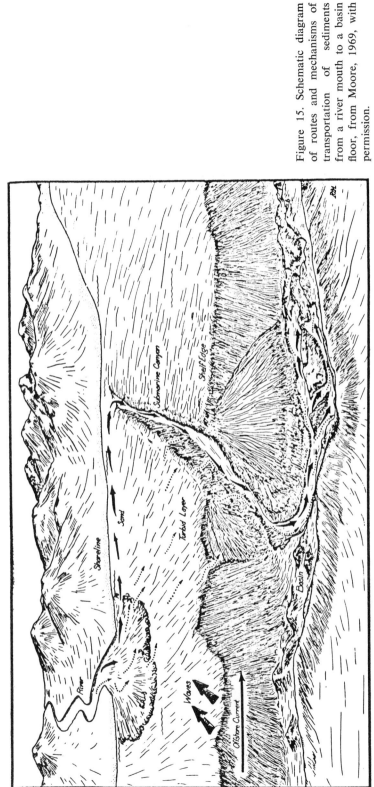

Figure 15. Schematic diagram of routes and mechanisms of transportation of sediments from a river mouth to a basin floor, from Moore, 1969, with permission.

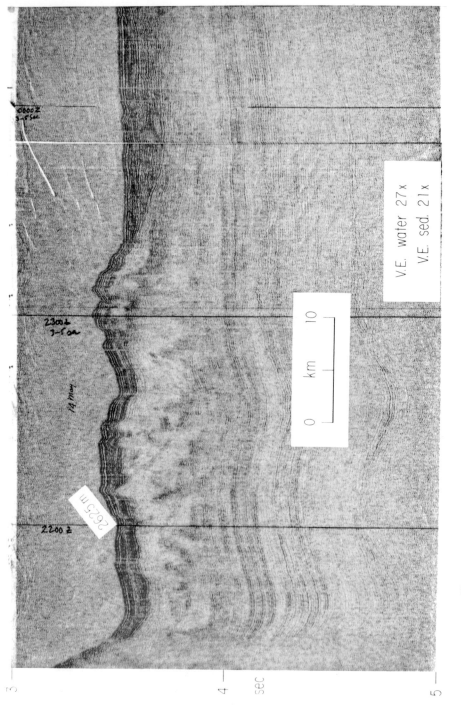

Figure 16. An "olistostromal" or slump mass lying on the surface of the eastern margin of the Bengal Fan, at the foot of the continental slope west of the Irawaddy Delta, Burma.

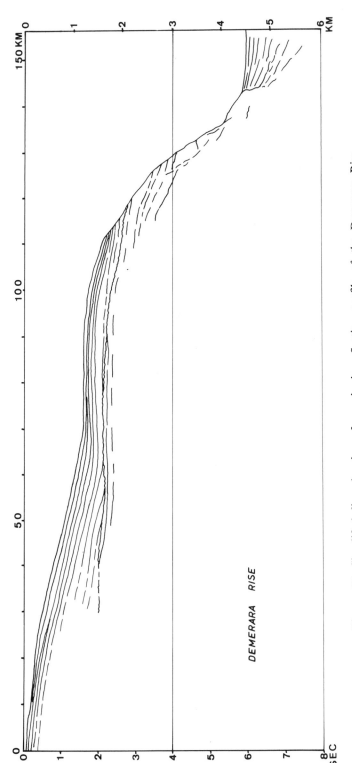

Figure 17. Simplified line drawing of a seismic reflection profile of the Demerara Rise marginal plateau, northeastern South America, showing a subsided late Cretaceous shelf with some late Tertiary and Quaternary sediments now prograding across the landward margin of the plateau.

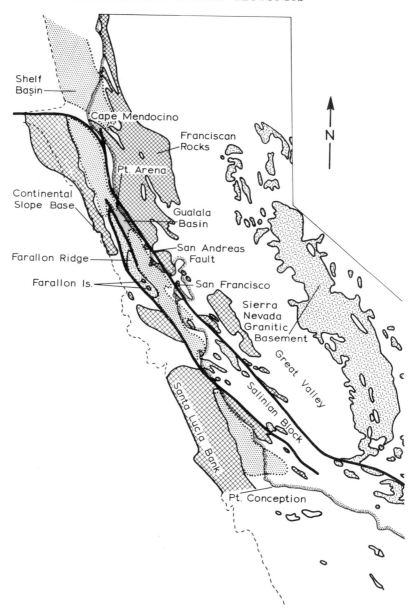

Figure 18. Filled late Cenozoic basins of the central California "continental border-land" shown in stippled pattern, and distribution of outcrop and shallow subcrop of basement rocks. These basement rock ridges on the continental margin were emplaced laterally during the late Tertiary by the San Andreas transform fault system. From Silver and others, 1971, with permission.

zones are particularly interesting geologically. They are generally formed as a result of scraping off of the sediments riding the oceanic plate into a Benioff underthrust, such as off northern California (Silver, 1971) (Fig. 19). The sediments thus scraped off are uplifted, deformed, and

INTERPRETATION OF REFLECTION PROFILES

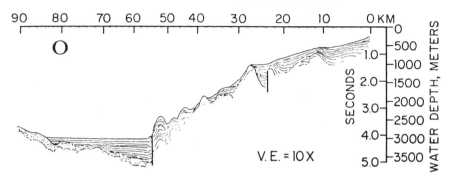

Figure 19. Line drawing from seismic reflection profile of filled basins, uplifted deeper water sediments forming the barriers, and the marginal plateau off northern California above the subduction zone of the Gorda plate. From Silver, 1971, with permission.

subsequently underthrust by younger sediments and ocean floor that originally lay farther seaward. This uplift and thrusting may create various kinds of basins. In the extreme, where the sediment column overlying the ocean floor is thick and rate of subduction is rapid, a sedimentary ridge above sea level may be formed, as in the Mentawai Islands off Sumatra and Nicobar and Andaman Islands (Fig. 20). The basin off Sumatra occupying the arc-trench gap is subsiding and has accumulated significant thicknesses of turbidites derived from the older arc on Sumatra, from the active volcanic arc, and from the sedimentary ridge. Similar basins and plateaus are formed on the landward walls of trenches or over subduction zones too flooded with sediments to have topographic trench depressions.

Base-of-Slope Environments

Those sediments that escape from all of the various traps and sites of deposition along the way from alluvial plains, the shore zone, the shelf and shelf basins, and the slope and slope basins, finally arrive at

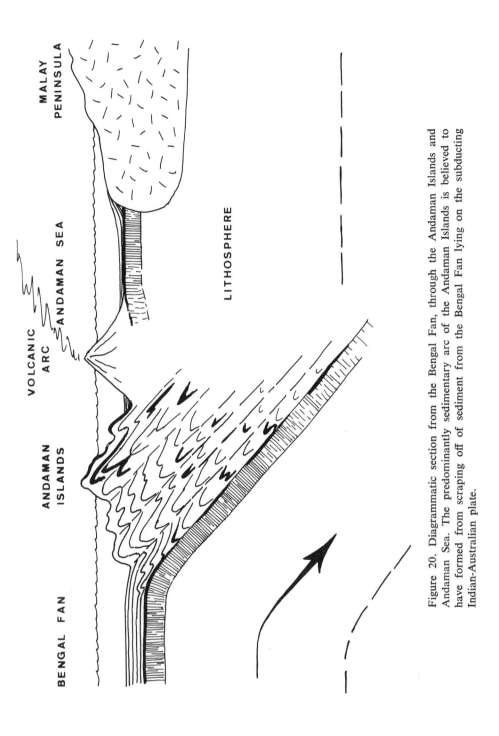

Figure 20. Diagrammatic section from the Bengal Fan, through the Andaman Islands and Andaman Sea. The predominantly sedimentary arc of the Andaman Islands is believed to have formed from scraping off of sediment from the Bengal Fan lying on the subducting Indian-Australian plate.

the ocean floor at the base of the continental slope. They may find on their arrival a marginal depression or deep sea trench, or they may find just a decrease in gradient analogous to the foot of a mountain slope. Either constitutes another possible sediment trap. Modes of transport to the base of the slope may be turbidity currents in canyon axes, turbid layer transport, or slumping or sliding from the slope, which may in turn generate turbidity currents. In addition, pelagic contribution is always present but is diluted to varying degrees by the terrigenous contribution, especially by turbidity currents.

Trench deposits vary widely in volume and nature, depending upon rate of subduction, rate of supply from land, and number and efficiency of sediment traps along the way. Coarse sediments, even gravel, boulders, and olistostromal masses (Fig. 16) of unlimited size are possible, but their relative contributions are related to Quaternary effects. Gravels and sands may be less common today than normal, while olistostromal masses may be relatively more common. Sources may be continental, volcanic from the island arc, reworked sediment from the landward slope, or even ultrabasic mantle rock from the landward slope. Transport may be transverse to the trench axis, but is probably more commonly longitudinal from a source at one end.

Midplate continental margins accumulate slope-base sediments as deep sea fans or continental rises. A deep sea fan (Fig. 21) is a turbidite body derived mainly from a single major source or feeder submarine canyon. Discrete separate fans lie off western North America and off many of the major river systems of the world such as the Amazon, Indus, and Ganges-Brahmaputra. The feeder submarine canyon is only the upper part of the drainage system, and may be eroded into preexisting rock of the shelf and slope or it may be formed by nondeposition of sediment along a canyon axis in a thick subsiding sedimentary continental terrace. At the base of the slope or apex of the fan, the turbidity currents leave the deep submarine canyon and flow through fan valleys (Shepard and Dill, 1966) like rivers down the surface of the fan, confined normally by their own natural levees (Fig. 21). They are known to meander and braid, and may migrate gradually or break through their levees and shift abruptly. Such channel migration can create very complex structures of natural levee deposits and filled channels piled one on top of another. Such complex structures are easily misinterpreted in seismic reflection records, and we believe it is possible that many of the "deformed" sections (Fig. 22) that have been interpreted as

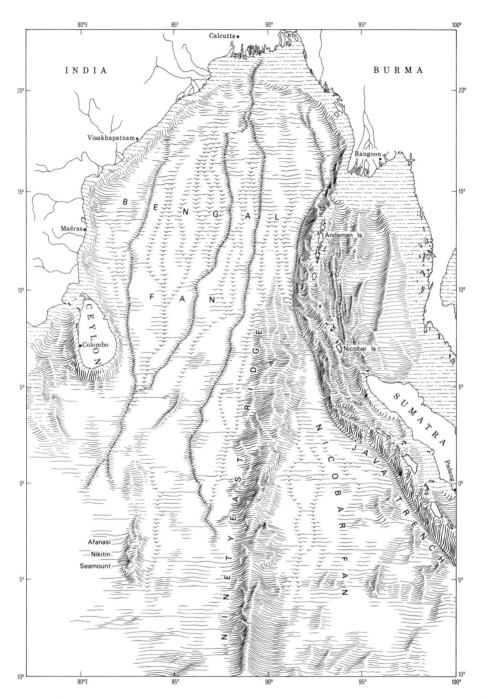

Figure 21. Physiographic diagram of the Bengal Fan, northeastern Indian Ocean, showing the Swatch of no Ground submarine canyon on the shelf off the Ganges-Brahmaputra delta and the network of turbidity current fan valleys running the length of the fan. From Moore, Emmel, and Curray, in preparation.

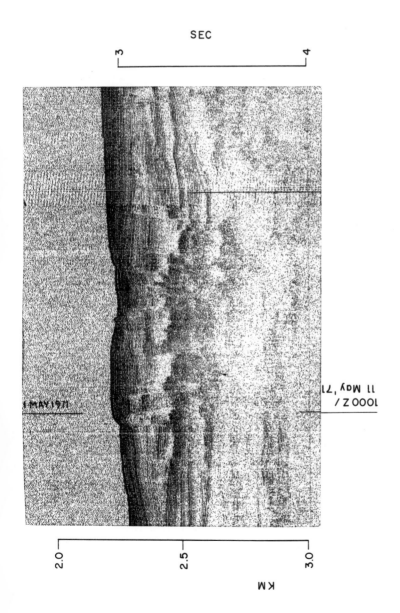
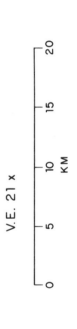

SEC

3

4

1 MAY 1971

11 May '71
1000 Z /

2.0

2.5

3.0

K M

V.E. 21 x

0   5   10   15   20

KM

Figure 22. Seismic reflection profile showing a complex of abandoned and filled fan valleys on the proximal part of the Bengal Fan.

slumped sediments in fans and continental rises may really be migrating channel complexes (Moore, Emmel, and Curray, in preparation).

Small turbidity currents are normally contained within the natural levees, and deposit their sediments in the channels. Larger flows may overtop the levees anywhere along the way and spread out laterally depositing sediments across the fan surface. Sediment distributions of fans have been thoroughly studied by piston cores (see, for example, Shepard et al., 1969; Normark and Piper, 1972), but the high rates of lowered sea level deposition have limited penetration to Holocene and latest Pleistocene. The rates of deposition must certainly be atypical of pre-Quaternary sediments of the fans because of the great changes in proximity of source during transgressions and regressions across the adjacent shelves. It is nevertheless safe to conclude that fan sediments are alternating muds and graded sands, with concentration of sands in and adjacent to the fan valley systems.

A continental rise (Fig. 2) is a complex of coalesced deep sea fans. Where the source is not a single point or submarine canyon, but is instead a row of submarine canyons, or essentially a line source, the discrete fans coalesce into a continental rise. Sediments are brought to the base of the slope mainly by turbidity currents. Channels like fan valleys traverse the slope, but meanders and channel migration have not been reported. Some of the sediment must pass on through the channel system to the abyssal plain, but some is deposited. Geostrophic currents ("contour currents," Heezen et al., 1966) then move some of the sediment laterally along the contour, to the south in the case of eastern United States. These geostrophic currents are intensified on the western sides of the oceans, but do occur elsewhere. Velocities of up to 20 cm/sec are reported at the sea floor on the continental rise off eastern United States.

The relative importance of turbidity current versus geostrophic current transport is controversial. Stanley et al. (1971) have suggested that contour currents are not as important as previously believed, and that the discrete fans can in fact be seen in detailed surveys.

The surface sediments of continental rises and fans are "equilibrium" or "modern" sediments, but the drainage systems are in many cases not yet in equilibrium with present flow regime. During Pleistocene lowered sea level, with the shoreline approximately at the shelf break, rate of supply to the fans and rises was higher. Fan valleys adjusted their cross sectional areas to accommodate this increased flow. Since the Holocene transgression, rates of discharge and size and frequency of turbidity currents have been reduced, and many of the fan valleys appear to be aggrading to adjust to the reduced flow. The principles of river aggradation or degradation, as well as meandering and migration can be applied (Mackin, 1948; Leopold et al., 1964).

Ocean Floor Environments

The sediments discussed in the previous section on slope-base environments are actually deposited on the ocean floor or oceanic crust, but are considered as part of the continental margin in the broad definition used here. Underlying the turbidites of fans, rises, and trenches and lying seaward of their distal ends are the pelagic sediments of the ocean, with their interstratified and underlying volcanics.

Pelagic or oceanic sediments are composed of mixtures of five different kinds of constituents: terrigenous, biogenic, authigenic, volcanic, and extra-terrestrial. Depending upon the relative proportions of these constituents, names are applied to the sediments such as red clay (which is actually brown), siliceous ooze, radiolarian ooze, carbonate ooze, foram ooze, etc. Turbidites may be interbedded or may bury the pelagic deposits as a fan, continental rise, or abyssal plain deposit progrades. Terrigenous contributions may also be carried long distances to midocean regions by wind transport and by geostrophic or other bottom currents. Transport in suspension in the water column for long distances is probably not volumetrically very important. Authigenic constituents are important where rates of accumulation and dilution by other constituents are low, and form the well-known phosphorite and manganese nodules and crust.

One of the important aspects of ocean floor sediments is their association with and contributions of volcanic materials. Ophiolites and the Steinmann Trinity (serpentinite, spilite, and radiolarite) have long been recognized as fragments of the ocean floor preserved in mountain ranges. Such associations in mantle rocks, altered pillow lavas and dikes, and pelagic sediments are generally interpreted today as having formed near "midocean" spreading ridges, by first deposition of oceanic sediments on the newly created sea floor. The interpretative section of Pacific abyssal hills made by Luyendyk (1970) based on detailed surveying and sampling bears a striking resemblance to sections of ophiolite bodies studied on land (Page, 1972), (Fig. 23).

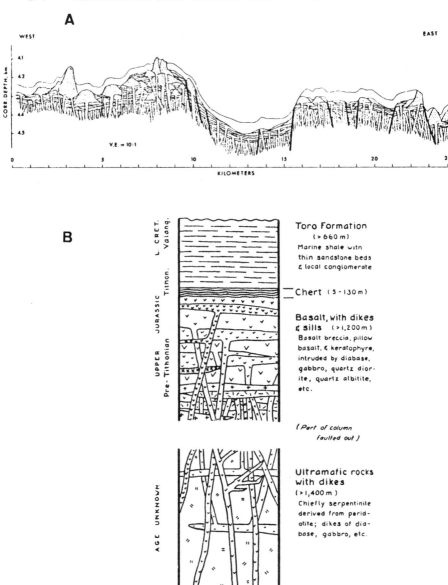

Figure 23. Comparison of: A. model section of Pacific abyssal hills (Luyendyk, 1970), as constructed from precision survey with a near-bottom geophysical package, coring, and dredging. B. schematic geologic column constructed from study of a remnant of the Mesozoic ocean floor preserved in the Coast Ranges of California (Page, 1972).

A wealth of new information on pelagic sediments is now available in the reports of the JOIDES Deep Sea Drilling Project.

Sediment Transport and Deposition in the Ocean

The following modes of transportation in the ocean contribute to the overall transport patterns of sediments.

1. Hintersurf mechanisms. This includes river flow, wind, and tidal current transport into and out of shore-zone environments.

2. Surf. Wave action in the surf zone is responsible for erosion and deposition of beaches and barriers and for longshore distribution of both bedload and suspended load.

3. Infrasurf bedload mechanism. This includes tidal current, rip current, wave and storm surge transport of bed load. Null point theory suggests that some sands may escape seaward from the surf zone, but the evidence seems to show that more sand is transported onshore from the shelf into the surf zone than offshore.

4. Suspension. Fine sediments are transported from river mouths to the inner continental shelf beyond the surf zone, but not much sediment is transported by suspension beyond the shelf.

5. Turbid layer. Fine sediments are temporarily deposited on the shelf and resuspended by long period wave surge, bottom currents, and organisms. Transport in the turbid layer is in response to currents and gravity, and carries sediment to the continental slope.

6. Turbidity current. Turbidity currents generated on steep slopes and especially in the heads and axes of submarine canyons can transport sediment of all sizes to all parts of the ocean that are downslope.

7. Geostrophic and other currents. Currents near the sea floor transport silts, clays, and even fine sands on continental rises and in "wind rows" on the deep sea floor.

8. Gravity mechanisms, sliding and slumping. Slopes over-steepened by tectonics or deposition, especially during the Pleistocene, may be unstable. Allochthonous or olistostromal masses of sediment may slide long distances down gentle slopes once set in motion.

9. Wind. Fine sediments are carried in large quantities to all parts of the oceans. This is probably the principal mechanism for transport of terrigenous sediment to the centers of large oceans rimmed with sediment traps.

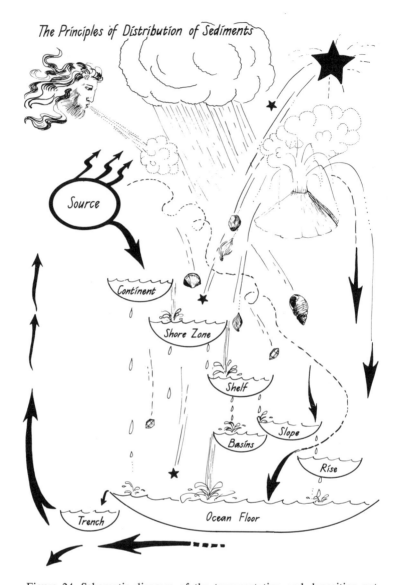

The Principles of Distribution of Sediments

Figure 24. Schematic diagram of the transportation and deposition pattern of sediment entering the ocean along a continental margin. Terrigenous sediment is derived from a source on land, and passes successively from one cascading pool (depositional environment) to another, seeking a permanent resting place. None of the pools or environments is perfect: all have leaks, and will retain only a portion of the sediment. Some goes into solution at the source or along the way, and may re-enter the system at any point as authigenic or biogenic material. Bypassing is common. A submarine canyon may, for example, head in the shore zone and funnel sediments directly to some of the deeper environments. In the meantime, the possibility of introduction of wind-blown, volcanic, or even extra-terrestrial material also exists at all levels. Tectonics may intervene anywhere along the way and uplift previously deposited material high enough, whether above or below sea level, to provide a new erosional source, or else subduct the sediments to be recycled completely.

10. Solution. Important for calcium carbonate, silica, evaporites, and other authigenic minerals.

11. Tectonic. Sea floor spreading carries sediments on the sea floor thousands of kilometers to subduction zones where they may be either carried down into the asthenosphere or scraped off, uplifted, and thrust into ridges or mountain ranges.

By these mechanisms, sediments entering the ocean are carried varying distances before being deposited. In passing from a river mouth, terrigenous sediment has many sediment traps lying in front of it. Deposition in any of these traps may be permanent, in a geological sense, or it may be temporary; and the sediment particle may then pass on to confront the next trap. Ultimately, some of the sediment may arrive on the ocean floor, but by far the greatest amount has been lost along the way to different depositional environments. This transport pattern is shown diagrammatically in Figure 24.

EFFECTS OF THE QUATERNARY

Some of the effects of Quaternary climatic and sea level fluctuations have been mentioned along with descriptions of the environments. These effects are summarized in Table II.

TABLE II
EFFECTS OF THE QUATERNARY

Shore Zone
 Sediments mostly "equilibrium."
 Environments started forming 3000 to 7000 BP.
 Environments evolving rapidly.
 Deltas young and restricted to inner shelf.
 More estuaries than usual.
 More barrier-lagoon systems than usual.
 Fewer tidal flats than usual?

Continental Shelf
 ca. 50% relict or reworked relict sediments.
 Morphology unusual:
  Most continental margins have shelves.
  Shelf breaks sharper than usual.
  Shelf breaks relatively uniform in depth.

TABLE II

EFFECTS OF THE QUATERNARY (*Continued*)

Continental Slope Environments

Sediments mostly "equilibrium."

Rates of deposition relatively low today.

Slopes more dissected than usual.

More submarine canyons than usual.

More slump and slide masses than usual.

Fewer coarse sediments being deposited than usual.

Base of Slope Environments

Sediments mostly "equilibrium."

Rates of deposition relatively low today.

Fewer coarse sediments than usual.

Larger fans and rises than usual.

Some fan valleys are incised.

Ocean Floor

Sediments mostly "equilibrium."

Rate of terrigenous supply relatively low.

Fewer coarse sediments than usual.

PLATE TECTONICS, GEOSYNCLINES, AND OROGENY

Environments of deposition of sediments are arranged in different combinations along different types of continental margins. Different types of continental margins are, in turn, to a first approximation, related to plate position. Collectively, the environments of a continental margin plus the adjacent ocean floor may, if sediment volumes are great, constitute an equivalent to geosynclines of the past. If one of our objectives in geology is paleogeography, or explanation of ancient sediments in terms of modern or at least "non-Quaternary" environments, then we must also attempt to relate the important complexes of environments of deposition to ancient basins or geosynclines.

For lack of a better or more convenient term, I will use *geosyncline* to designate a thick, sometimes elongate, large volume of sediment and associated intrusives and extrusives. Such accumulations of sediments can be formed in two different ways. First, they can accumulate by deposition in a subsiding environment or a complex of subsiding environments adjacent to a large and long-continued source of sediment. This must require a continental margin in the broad sense used here in order to have both the source and the sink. Second, a significant accumulation of sediments may be concentrated tectonically by a snowplow effect. One lithospheric plate overriding another may accumulate a great volume of sediment in front of it by scraping off the veneer of sediment and some of the oceanic crust passing beneath it in subduction. The result may be great cumulative thicknesses of sediment and associated ocean floor rocks, not as depositional sequences, but as mélanges (Hsü 1968).

A geosynclinal depositional accumulation of sediment does not necessarily make a mountain range. It is true that the subduction method of mélange accumulation deforms the sediment as it accumulates, intrudes it and mixes in some volcanics, but how about the depositional accumulations? Many models have recently been proposed for making mountain ranges of sedimentary accumulations. In proposing such models, however, we must heed the advice of Coney (1970) and abandon the prejudice of geotectonic cycles. The tectonics of creating a "geosynclinal" accumulation of sediment can be quite independent of the tectonics of making the accumulation of sediments into a mountain range. Geosynclines are related to a "determining background" (Krynine, 1951), or the creation of a source, transport paths and mechanisms, and a complex of subsiding traps. Destruction of that pile of sediment may result not from a pre-determined sequence of events in a predictable geotectonic cycle, but from a sequence of random events of plate tectonics.

Most of the literature on geosynclines prior to about 1969 had a very clear objective: to interpret ancient geosynclines, sometimes also searching the modern world for equivalents. Many such efforts to find modern equivalents have been unsuccessful, and seem to conclude that the great sediment accumulations we find today are significantly different from ancient geosynclines. Special names have been proposed for some of these modern sediment accumulations, although they are sometimes simply dismissed from consideration as equivalents. I believe that this approach is fundamentally wrong. I believe that great accumulations of sediments, what I am with my loose terminology calling *geosynclines*, are permanent monuments. Such accumulations are probably rarely totally destroyed even during long subsequent geological time. Some parts of these sediment accumulations may be subducted; other parts are destroyed by erosion; but some part remains. It may be deformed and metamorphosed in the core of a mountain range, or it may even remain undeformed and buried for long periods of geological time, but the chances of total destruction are slight. We must, therefore, account for all types of great accumulations of sediment, ancient or modern, in discussing and classifying geosynclines.

Most classifications of geosynclines are provincial, both in time and space. Geosynclinal models and geotectonic cycles are proposed generally using the Alpine and Appalachian geosynclines as type examples. Other geosynclines, being different, then require different terminology and sequences and deviate from the primary models. Until the past few years, all such models were based on interpretations of these ancient sediment accumulations, both with complex subsequent histories, so that even the models are changed frequently with new work, interpretations, and concepts.

All mountain ranges are different; all continental margins are different; and all geosynclines are and have been different. We can find patterns and similarities, and can group and classify, but we should not expect the basic simplicity that has sometimes been assumed in the past. Great benefit and advances in geological thinking have been derived from past work on geosynclines, but today we should reconsider all of that in the light of newer discoveries in both terrestrial and marine geology and geophysics and in the terms of the newer concepts of tectonics.

Geosynclinal models should be set up from what we now know about the modern oceans. All possibilities should be considered in the belief that all environments and combinations of environments that exist today must have existed at some time in the past, and many that have existed in the past are not represented in the modern world. Although some possible combinations of environments and tectonic settings may be considered less likely, we should try to maintain as many models and working hypotheses as possible. And, finally, we should consider all of the possible subsequent tectonic fates of these geosynclinal accumulations of sediment in terms of our present understanding of plate tectonics.

Many discussions of possible actualistic geosynclines and their orogenic fates have been published. The possible combinations are limited, and the similarities among the published discussions are obvious. The discussion to follow will, therefore, contain many of the elements already well described and discussed in papers by Dietz (1963), Dietz and Holden (1966), Wilson (1968), Mitchell and Reading, (1969, 1971),

Dewey (1969), Dewey and Horsefield (1970), Dewey and Bird (1970), Dickenson (1971 a,b,c,d), and many others. An attempt is made here to include some elements not previously discussed by other authors in the belief that almost all combinations are possible in the geologic past, and that we tend to try to fit these complex past situations into too few actualistic models or working hypotheses.

Geosynclinal Accumulations of Sediment

One method of evaluating as many potential modern "geosynclines" as possible is to construct a matrix of the important elements of Table I. This has been done in Table III, using the four different plate positions of continental margins plus the island arc–deep sea trench complex of the oceanic environments. Twenty possibilities result, including these elements alone as one-sided geosynclines or in combinations with each other as two-sided geosynclines. Eight of these will be considered as the more important possible modern geosynclines, although they are certainly not of equal importance. Some other combinations that involve the bringing together of different elements by subduction will be considered later in the discussion of possible orogenies.

The eight geosynclines to consider are of three types: passive, midplate, or *epeirogenic* geosynclines; *tectogenic* plate-edge geosynclines; and *orogenic* plate-edge geosynclines. The latter might be termed autoorogenic or synorogenic, capable of producing mountain ranges of the sediment as it accumulates, without a change in plate arrangement, relative motion, or collision.

Purists in the use of the classical terms *epeirogenic, tectogenic,* and *orogenic* may not accept my usage. I apologize to those whom I may offend, but prefer to use simplified and somewhat modified meanings of these terms rather than to introduce new terms. I will use these terms as defined in the AGI Dictionary (AGI, 1962), but I refer the reader to the excellent discussions in Aubouin (1965).

Epeirogeny. "The broad movements of uplift and subsidence which affect the whole or large portions of continental areas or of the ocean basins." (ibid., p. 162). I will equate this generally with midplate position, whether it is continental, oceanic, or continental margin.

Tectogenesis. "The process by which rocks are deformed; more specifically, the formation of folds, faults, joints, and cleavage. Orogenesis, a term often used for these processes, should be restricted

TABLE III

ACTUALISTIC GEOSYNCLINE MODELS

	MIDPLATE CONTINENTAL MARGIN	SUBDUCTION CONTINENTAL MARGIN	TRANSFORM FAULT TYPE CONTINENTAL MARGIN	NASCENT CONTINENTAL MARGIN	ISLAND ARC AND TRENCH COMPLEX
ELEMENT OCCURRING ALONE	*ATLANTIC TYPE*	*ANDEAN TYPE*	*CALIFORNIA TYPE*		*ISLAND ARC AND TRENCH COMPLEX*
COMBINATION WITH ISLAND ARC AND TRENCH	*JAPAN SEA TYPE*	collision[1] type C_sI_+	collision type $I\ C_t$		collision types I_+I_+
COMBINATION WITH NASCENT CONTINENTAL MARGIN				*NORTHERN GULF OF CALIFORNIA TYPE*	
COMBINATION WITH TRANSFORM CONTINENTAL MARGIN	*SMALL TRANSFORM OCEAN BASIN TYPE*	collision type C_sC_t			
COMBINATION WITH SUBDUCTION CONTINENTAL MARGIN	collision type C_sC_m	collision type C_sC_s			
COMBINATION WITH MIDPLATE CONTINENTAL MARGIN	*ABORTED OCEAN TYPE*				

[1] Collision types and the abbreviations used are explained in the text and Table IV.

to processes resulting in morphological features." (ibid., p. 492). I will assume that tectogenesis is generally restricted to plate edges, realizing that plate edges are not always sharp, but are sometimes broad zones, especially when located within continental crust.

Orogeny. "The process of forming mountains, particularly by folding and thrusting." (ibid., p. 358). I am assuming, as have many others (see, for example, Ahmad, 1968, plus the authors cited for discussions of actualistic geosynclines and orogenies) that orogeny occurs only along plate edges with some component of convergence. Tectogenesis accompanies orogeny here.

A. EPEIROGENIC, MIDPLATE, OR PASSIVE GEOSYNCLINES

1. ATLANTIC TYPE GEOSYNCLINE, shelf-rise type, or midplate continental margin type geosyncline (Fig. 25 A1). Examples are many and well known, such as the east and gulf coasts of the United States, the Bay of Bengal Ganges-Brahmaputra delta and shelf and the adjacent Bengal deep sea fan, etc. With time and continued spreading, a continental rift valley evolves to an incipient or nascent ocean, and then to two midplate continental margins. These may be considered geosynclines at any step of this evolution with sufficient volume of sediment. Subsequent orogenesis may occur only by rearrangement of plates (Act.), or by collision with a subduction zone (C_sC_m, IC_m). These orogenies are discussed and explained in the next section and listed in Table V, with the abbreviations used here.

This type of geosyncline has been compared with ancient orthogeosynclinal coupled mio- and eugeosynclines by Drake et al. (1959), Dietz (1963), etc. Kay (1951) proposed the name *paraliageosyncline* for midplate continental margins without much associated volcanic rock, but did not cite ancient examples that have been incorporated into mountain ranges. Many authors dispute the comparison of continental rise with eugeosyncline because of the apparent absence of volcanics.

2. ABORTED OCEAN TYPE GEOSYNCLINE (Fig. 25 A2). During the evolution from rift valley to Atlantic type ocean, spreading may stop at any time. Aborted rift valleys have been delineated in Africa (Burke et al., 1971). If spreading stops after a narrow ocean has opened, a passive, midplate, two-sided basin is produced, which if sediment volumes are sufficient may be considered a type of geosyncline. Possible examples are Baffin Bay and Davis Strait between Greenland and

ACTUALISTIC   GEOSYNCLINE   MODELS

A. EPEIROGENIC TYPES

1.   ATLANTIC

2.   ABORTED OCEAN

3.   JAPAN SEA

B. TECTOGENIC TYPES

1. CALIFORNIA

2. TRANSFORM OCEAN

3. N. GULF OF CALIFORNIA

C.   OROGENIC TYPES

1. ANDEAN

2. ISLAND ARC

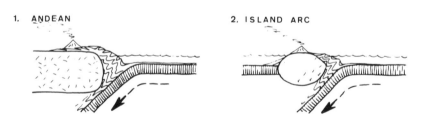

Figure 25. Diagrammatic sections of actualistic geosyncline models. In types B 1 and B 2, the circled cross signifies transform movement away from the observer, the circled dot signifies transform movement toward the observer.

Canada, the whole of the Gulf of Mexico, and Mozambique Channel. Subsequent orogenesis can result only from rearrangement of plate boundaries (Act.) or possibly by end-on collision with a subduction zone (C_sC_m, IC_m). Such geosynclines may have a fair chance of complete filling with sediment and preservation as a part of a craton for long periods of geological time.

3. JAPAN SEA TYPE GEOSYNCLINE (Fig. 25 A3) (Mitchell and Reading, 1969). This type is formed by location of an island arc complex subduction zone a short distance seaward of a midplate continental margin. The outer part of the island arc complex is orogenic, in the terms used here, but the basin itself, located between the island arc and the continental margin, is midplate, hence its inclusion here in the classification. Karig (1970) has shown that some such basins are regions of high heat flow and irregular spreading so tectogenesis and volcanism may occur contemporaneously with deposition. Partial or complete orogenesis may result again from rearrangement of plates or changes in relative motion or by various collisions (Table V).

B. TECTOGENIC, PLATE-EDGE, GEOSYNCLINES

1. CALIFORNIA TYPE GEOSYNCLINE, or transform fault zone continental margin type (Fig. 25 B1). The continental margins off central, southern, and Baja California and northern Venezuela are situated generally along present or past transform fault plate edges. Tectogenic deformation, strike-slip faulting, and basin formation occur, but not orogenesis. Very considerable volumes of sediments may be accumulated in the elongate basins created by strike-slip fault emplacement of tectonic ridges or sediment dams (Fig. 14) and filled basins may result (Fig. 18) if a source is available. Subsequent orogenesis may result from collision with a subduction zone (C_sC_t, IC_t), by a change in relative plate motion to add a significant component of convergence or subduction (Chg.+), or by migration of a triple junction along the transform fault to put a different plate with a component of subduction in contact with the continental margin (C_tR) (Atwater, 1970).

2. SMALL TRANSFORM OCEAN BASIN GEOSYNCLINE (Fig. 25 B2). Many arrangements of small ocean basins are possible. The combination of a midplate continental margin facing a transform fault continental margin across a small ocean basin is included here because of the possibility that the Alpine-Mediterranean portion of Tethys had this configu-

ration from Middle Jurassic to Late Eocene (see, for example, Hsü, 1971). The relative motions of the smaller plates and continental fragments were certainly vastly more complex, but there was perhaps relative lateral motion first in a left lateral sense and then in a right lateral sense between Africa and Europe, before Alpine convergence and subduction. This group in the classification is a catch-all for many complex possibilities involving relative lateral motion. Orogeny (Table V) probably results from addition of a subduction component.

3. NORTHERN GULF OF CALIFORNIA TYPE GEOSYNCLINE, or nascent ocean basin geosyncline (Fig. 25 B3). The Gulf of California is a young ocean basin being formed by relative separation between the North American and Pacific plates with a complex set of short spreading rifts and transform faults. The southern part of the Gulf is clearly oceanic crust (Phillips, 1964), but, because of the tremendous sediment load entering the northern sector, it has an intermediate type crust. Moore (1973) has introduced the concept of "clastic compensation" to cover the combined effects of plate separation and simultaneous sedimentary burial, which preclude the formation of normal oceanic crust and extrusion of lavas. In most plate configurations, this would be an ephemeral feature, which would evolve into a normal ocean basin and Atlantic type continental margins as separation increases. Because of the oblique opening of the Gulf, however, projection of present plate motions shows that it will not become significantly wider, and could continue to fill for 50 MY or more before the continental fragment on the Pacific plate enters the Aleutian Trench. Earlier orogeny could occur by any change in relative plate motion to add a component of convergence between the two plates involved (Act, or Chg. +), or a stop in the relative plate motion would convert this into an Aborted Ocean Type Geosyncline.

C. OROGENIC, PLATE-EDGE GEOSYNCLINES

Only two possibilities will be listed and discussed here, a continental margin subduction zone and an island arc complex subduction zone. Other possibilities are the many combinations involving subduction zones. These are two-sided geosynclines, but the two sides are not of much importance to each other until they come into proximity and collision commences. Prior to that time, they are independent single elements. Collision events will be discussed in the next section on Orogenies.

1. ANDEAN TYPE GEOSYNCLINE, or subduction continental margin (Fig. 25 C1). Sediment is accumulated both by deposition and by lateral accumulation by the snowplow effect. Orogeny accompanies accumulation. Some sediment may be subducted to be metamorphosed and subsequently reappear by isostatic uplift after termination of subduction, some sediment is scraped off the subducting plate and uplifted in a series of thrust sheets and folds to form a sedimentary ridge (Fig. 20). Interarc and arc-rear basins are created and accumulate sediments and volcanics, while andesitic volcanism occurs at a distance behind the subduction zone corresponding to a Benioff plane depth of 175 ± 100 km (Dickenson, 1971a). Many different collision events are possible (Tables IV and V).

2. ISLAND ARC COMPLEX GEOSYNCLINE (Fig. 25 C2). Much of what has already been said for Andean Type Geosynclines applies here. In a purely oceanic island arc, sediment sources may not be great, and most will never accumulate sufficient sediment to become geosynclines. They will accumulate large volumes of sediment only if it is derived from elsewhere, in general from a continental margin source. This will occur only during the various stages of collision or near collision, as a continental margin moves toward or alongside a trench. An example of the latter is the distal portion of the Bengal Fan, the outermost part of a midplate continental margin, which is being carried obliquely into the Java–Sumatra Trench subduction zone (Figs. 20 and 21). Full-scale collision will never occur along this subduction zone between continent and island arc, but the large volume of terrigenous sediments of the fan is nevertheless being carried by and scraped off the subducting plate.

Either continental margin or island arc subduction zone geosynclines may experience changes in tectonic style by changes in relative plate motion, by collisions, or by cessation of convergence of the plates. These changes will be discussed in the next section and listed in Tables IV and V.

Orogenies

Geosynclinal accumulations of sediments may meet with a large number of possible fates by the random events of plate tectonics. While some are certainly more probable than others, there is no single predestined set of events or geotectonic cycle in the evolution of a given type of geosyncline. Rather than setting up models from our interpretations of

OROGENIC   CHANGES

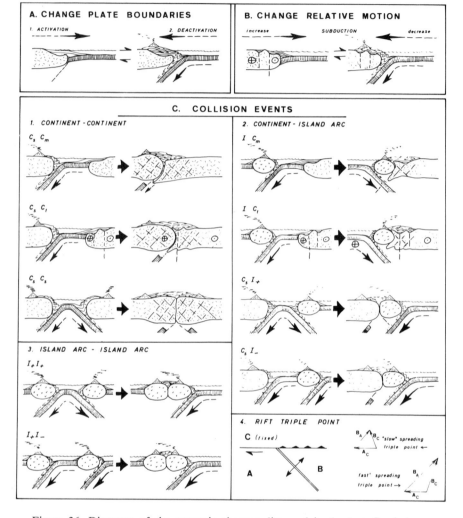

Figure 26. Diagrams of the orogenic changes discussed in the text. Symbols are the same as Figure 25. Abbreviations C_sC_m, etc., are keyed with the text. Note the indications of active and inactive volcanism. Section C4 is a triple junction in plan view, showing a spreading rift with a double line, a transform fault with arrows indicating relative movement, and a subduction zone with barbs on the hanging or overthrust wall. Plate C is considered fixed in the frame of reference. Vector diagrams represent the possibilities discussed in the text. B_A indicates the motion of plate B relative to plate A, etc. The upper possibility represents "slow" spreading across the rift relative to the other plate motions, so that the triple point migrates to the left along the edge of plate C. If this is a continental margin, it is then converted from a transform type to a subduction type. The lower possibility represents "fast" spreading, and the reverse changes.

histories of the better known ancient geosynclines, we should consider all possibilities in the light of our present understanding of the concepts of plate tectonics. Orogeny must require at least some component of convergence or subduction, but other possible tectonic changes will be considered as well. These orogenies are tabulated with the geosynclinal types that they might affect in Table V, and they are illustrated diagrammatically in Figure 26.

A. CHANGE IN PLATE BOUNDARIES

A change in plate boundaries may occur by breaking up one plate into two or more smaller plates, by a shift in a plate boundary, or by cementing two or more plates into a single plate. Building upon previous published terminology (Dickenson, 1971c), I will call these processes "activation" and "deactivation."

1. ACTIVATION (Act.) is the breaking of one plate into two or more smaller plates, and thus creating new plate edges. This could occur in any of the geosynclinal types discussed, but would have the most significant impact in the case of one of the midplate or epeirogenic geosynclines. Dietz (1963), Hamilton (1969), Dewey (1969), Dickenson (1971c), and many others have discussed this possibility for Atlantic Type Geosynclines, and conversion of this type continental margin into the Andean Type (Fig. 26 A). Also of importance would be activation in either an Aborted Ocean Type or in a Japan Sea Type by shifting of a plate boundary.

2. DEACTIVATION (Deact.) (Fig. 26 A), is the cementing together of two or more plates. This could produce not the beginning of orogeny or tectogenesis, but the termination. Any kind of relative plate motion can stop, so any kind of plate edge geosyncline could thus be affected.

B. CHANGE IN RELATIVE PLATE MOTION (Chg. ±) (Fig. 26 B)

This can affect any of the plate edge geosynclines, but orogeny will result only in addition of or increase in the component of convergence. Changes can occur either in relative directions or speeds. Chg. + will designate addition of or increase in the component of subduction; Chg. — will designate a decrease.

C. COLLISION EVENTS

Many different collision events are possible and are occurring at all times in plate tectonics, such as plate against plate, seamounts into trenches, fracture zones into trenches, etc., but orogenically the significant collision events involve continental margins, island arc complexes, and other oceanic plate margins. The first two may result in mountain ranges and rearrangement of plate configuration.

TABLE IV

COLLISION EVENTS

		R	I	C_s	C_t	C_m
MIDPLATE CONTINENTAL MARGIN	C_m		$I\ C_m$	$C_s C_m$		
TRANSFORM FAULT CONTINENTAL MARGIN	C_t	$C_t R$	$I\ C_t$	$C_s C_t$		
SUBDUCTION ZONE CONTINENTAL MARGIN	C_s	$C_s R$	$C_s I_+$ $C_s I_-$	$C_s C_s$		
ISLAND ARC—TRENCH COMPLEX[1]	I	$T\ R$ $I\ R$	$I_+ I_+$ $I_+ I_-$			
OCEANIC RIFT SYSTEM: RIDGE AND TRANSFORMS[2]	R					

[1] I_+ designates an island arc complex subduction zone involved in a head-on collision. I_- designates one involved in a rear end collision.

[2] T designates an oceanic transform fault, to be considered here only for the special case of triple junction migrations, as explained in the text and in Figure 26.

Most of the simplest possibilities may be evaluated with a matrix of the various "collision elements" (Table IV). The elements considered include the different plate position continental margins, island arc complexes, and oceanic rift and transform systems. The latter includes the transform faults associated with the rifts, so that "collision" with an oceanic rift system implies creation of a triple point, which might then evolve into two triple points and contact between plates previously not in contact.

1. CONTINENT-CONTINENT COLLISIONS (Table IV, Fig. 26 C1).

   C_sC_m Andean type continental margin or geosyncline attempting to subduct an Atlantic type.

   C_sC_t Andean type attempting to subduct a transform type.

   C_sC_s Collision between two Andean types.

2. CONTINENT-ISLAND ARC COLLISION (Fig. 26 C2).

   I C_m Island arc-trench complex attempting to subduct an Atlantic type continental margin or geosyncline.

   I C_t Island arc complex attempting to subduct a transform type continental margin.

   C_sI_+ Collision face to face between a subduction continental margin and an Island arc.

   C_sI_- Collision between a subduction continental margin catching up with and overtaking an island arc complex from the rear.

3. ISLAND ARC-ISLAND ARC COLLISION (Fig. 26 C3).

   I_+I_+ Face to face collision between two island arc complexes.

   I_+I_- One island arc complex overtaking another from the rear.

4. OCEANIC RIFT SYSTEM "COLLISIONS" OR ENCOUNTERS (Fig. 26 C4.)

These encounters involve the geometric game of evolution of triple junctions, points where three plates meet. The rules are well explained and illustrated by McKenzie and Morgan (1969) and Atwater (1970). Triple junctions of importance here are those involving the intersection between an oceanic rift system, with or without transform faults, and continental margin geosyncline types, such as the Andean and California types, or the intersection between a rift and an island arc system. Such a triple junction is shown diagrammatically in plan view in Figure 26 C4. The right-hand side of the continental margin or island arc system has a component of subduction, while, to the left of the triple junction, pure transform motion is occurring. Depending in the two cases illustrated by vector plots on the relative velocities of spreading and transform motion, the triple point will migrate to the left or right, converting

Table V
Geosynclines and Their Possible Orogenies[1]

TYPE OF ACTUALISTIC GEOSYNCLINE		Act.	Deact.	Chg. +	Chg. −	C_sC_m	C_sC_t
ATLANTIC	C_m	C_s			yes		
ABORTED OCEAN	AO	NGC, STO, or C_sC_m			yes? "end-on"		
JAPAN SEA	JS	C_sI_-, STO, or $I C_m$	C_m	yes	yes		
CALIFORNIA	C_t		C_m	C_s		yes	
SMALL TRANSFORM OCEAN	STO	C_sC_t	AO	C_sC_m			
NORTHERN GULF OF CALIFORNIA	NGC		AO	C_sC_m, or C_sC_s			
ANDEAN	C_s		C_m	yes	C_sC_t, or C_m	yes	yes
ISLAND ARC	I		yes	yes	yes		

RANGE

TYPE OF TECTONIC OR OROGEN	$C_\times$							
$I\,C_m$	yes? "end-on"							yes
$I\,C_t$	yes		yes					yes
$C_\times I_+$	yes						yes	yes
$C_s I_-$							yes	yes
$I_+ I_+$	yes							yes
$I_+ I_-$	yes							yes
$C_t R$				yes				
$C_\times R$				C_s			C_t	
$T\,R$	yes, in reverse							yes, in reverse
$I\,R$	yes							yes

[1] Orogenies to which the different geosynclinal types might be subjected are indicated, either by "yes" or by the designation for the different geosynclinal type or orogeny that will result. Abbreviations for geosynclines are indicated above; abbreviations for orogenies are from the text or Table IV.

a possible geosyncline from Andean type to California type, or vice versa.

C_tR A continental margin and rift triple point tectonic change from California type (transform) to Andean type (subduction), or triple point migrating left (Fig. 26 C4).

C_sR The reverse change, from Andean to California type, or triple point migrating to the right.

T R Island arc and rift triple point migrating to the left, adding a component of subduction to a former transform segment.

I R Island arc and rift triple point migrating to the right, and changing a trench plate boundary into a transform boundary.

Discussion

We have reviewed eight different possible actualistic models for geosynclines. Some may have been common and important during the geological past, just as they are now, while others may have been rather rare. It would seem unlikely, however, that any of these tectonic settings are unique to the modern oceans: all of them must have occurred at some time during the long past. It also seems improbable that other types could have occurred in the past that cannot be predicted in terms of present plate tectonics models, unless they occurred before the lithosphere of the earth attained its present status and plate tectonic processes. I may, however, have missed some possible actualistic models in my survey.

These eight types may suffer a large number of possible orogenic histories. Just the simplest possibilities are enumerated above and shown in Table V. An almost infinite number of other possibilities exists by different combinations, through multiple plate configuration changes, or by bringing different geosynclinal types together on converging plate edges. By our present concepts, we should not expect any simple geotectonic cycles. All geosynclines and their subsequent orogenic history must have been different throughout geological time. As many different models as possible must be considered in attempting to evaluate them.

References

Ahmad, F., 1968. Orogeny, geosynclines and continental drift, *Tectonophysics,* v. 5, pp. 177–189.

American Geological Institute, 1962. *Dictionary of Geological Terms,* New York, Dolphin Books, 545 pp.

Atwater, Tanya, 1970. Implications of plate tectonics for the Cenozoic tectonic evolution of western North America, *Geol. Soc. America Bull.,* v. 81, pp. 3513–3536.

Aubouin, J., 1965. *Geosynclines,* Amsterdam, Elsevier Publishing Co., 335 pp.

Belderson, R. H., N. H. Kenyon, and A. H. Stride, 1971. Holocene sediments on the continental shelf west of the British Isles. In F. M. Delany, Ed., *The Geology of the East Atlantic Continental Margin,* Institute of Geological Sciences Report No. 74/14, pp. 157–170.

Bernard, H. A., 1965. A resumé of river delta types, Abstr., *Am. Assoc. Petroleum Geologists Bull.,* v. 49, pp. 334–335.

Bloch, M. R., 1965. A hypothesis for the change of ocean levels depending on the albedo of the polar ice caps, *Paleogeography, Paleoclimatology, Paleoecology,* v. 1, pp. 127–142.

Bloom, A. L., 1970. Paludal stratigraphy of Truk, Ponape, and Kusaie, Eastern Caroline Islands, *Geol. Soc. America Bull.,* v. 81, pp. 1895–1904.

Burk, C. A., 1968. Buried ridges within continental margins, *New York Acad. Sci. Trans.,* ser. 2, v. 30, pp. 397–409.

Burke, K., T. F. J. Dessauvagie, and A. J. Whiteman, 1971. The opening of the Gulf of Guinea and the geological history of the Benue depression and the Niger delta, *Nature Physical Science,* v. 233, pp. 51–55.

Coleman, J. M., and W. G. Smith, 1964. Late recent rise of sea level, *Geol. Soc. America Bull.,* v. 75, pp. 833–840.

Coney, P. J., 1970. The geotectonic cycle and the new global tectonics, *Geol. Soc. America Bull.,* v. 81, pp. 739–748.

Cromwell, J. E., 1971. Barrier coast distribution: a world-wide survey, Abstract Volume Second National Coastal and Shallow Water Research Conf., O.N.R., Washington, D.C., p. 50.

Curray, J. R., 1965a. Late Quaternary history, continental shelves of the United States. In H. E. Wright, Jr. and D. G. Frey, Ed., *The Quaternary of the United States,* Princeton University Press, pp. 723–735.

———, 1965b, Structure of the continental margin off central California, *New York Acad. Sciences Trans.,* ser. 2, v. 27, pp. 794–801.

———, 1969. Shore zone sand bodies: barriers, cheniers, and beach ridges. In D. J. Stanley, Ed., *The New Concepts of Continental Margin Sedimentation,* American Geological Institute, Washington, D.C., lecture no. 2.

Curray, J. R., F. J. Emmel, and P. J. Crampton, 1969. Holocene history of a strand plain lagoonal coast, Nayarit, Mexico. In A. Ayala-Casteñares and F. B. Phleger, Ed., *Lagunas Costeras, Un Simposio,* Universidad Nacional Autonoma de Mexico, pp. 63–100.

Dewey, J. F., 1969. Continental margins: a model for conversion of Atlantic type to Andean type, *Earth and Planetary Science Letters,* v. 6, pp. 189–197.

Dewey, J. F., and J. M. Bird, 1970. Mountain belts and the new global tectonics, *Jour. Geophysical Research,* v. 75, pp. 2625–2647.

Dewey, J. F., and Brenda Horsfield, 1970. Plate tectonics, orogeny and continental growth, *Nature,* v. 225, pp. 521–525.

Dickinson, W. R., 1971a. Clastic sedimentary sequences deposited in shelf, slope, and trough settings between magmatic arcs and associated trenches, *Pacific Geology,* v. 3, pp. 15–30.

———, 1971b. Plate tectonic models of geosynclines, *Earth and Planetary Science Letters,* v. 10, pp. 165–174.

———, 1971c. Plate tectonic models for orogeny at continental margins, *Nature,* v. 232, pp. 41–42.

———, 1971d. Plate tectonics in geologic history, *Science,* v. 174, pp. 107–113.

Dietz, R. S., 1963. Collapsing continental rises: an actualistic concept of geosynclines and mountain building, *Jour. Geology,* v. 71, pp. 314–333.

Dietz, R. S., and J. C. Holden, 1966. Miogeoclines (miogeosynclines) in space and time, *Jour. Geology,* v. 74, pp. 566–583.

Dott, R. H., Jr., 1963. Dynamics of subaqueous gravity depositional processes, *Am. Assoc. Petroleum Geologists Bull.,* v. 47, pp. 104–128.

Drake, C. L., M. Ewing, and G. H. Sutton, 1959. Continental margins and geosynclines: the east coast of North America north of Cape Hatteras. In *Physics and Chemistry of the Earth,* London, Pergamon Press, v. 3, pp. 110–198.

Emery, K. O., 1952. Continental shelf sediments of southern California, *Geol. Soc. America Bull.,* v. 63, pp. 1105–1108.

———, 1960. *The Sea Off Southern California: The Modern Habitat of Petroleum,* New York, John Wiley and Sons, 366 pp.

———, 1968. Relict sediments on continental shelves of world, *Am. Assoc. Petroleum Geologists Bull.,* v. 52, pp. 445–464.

———, 1969. The continental shelves, *Scientific American,* v. 221, pp. 107–122.

Emery, K. O., E. Uchupi, J. D. Phillips, C. O. Bowin, E. T. Bunce, and S. T. Knott, 1970. Continental rise off eastern North America, *Am. Assoc. Petroleum Geologists Bull.,* v. pp. 44–108.

Emiliani, Cesari, 1972. Paleotemperatures and the duration of the high temperature intervals, *Science,* v. 178, pp. 398–401.

Fairbridge, R. W., 1960. The changing level of the sea, *Scientific American,* v. 202, pp. 70–79.

———, 1961. Eustatic changes in sea level, *Physics and Chemistry of the Earth,* v. 4, pp. 99–185.

———, 1970. A review of: The new concepts of continental margin sedimentation—A.G.I. Short Course Lecture Notes, Philadelphia, 7–9 November, 1969. D. J. Stanley (Convener): Sedimentology, v. 14, pp. 337–346.

Fujii, S., and N. Fuji, 1967. Postglacial sea level in the Japanese Islands, *Jour. of Geoscience,* Osaka City Univ., v. 10, pp. 43–51.

Gould, H. R., and R. H. Stewart, 1955. Continental terrace sediments in the northeastern Gulf of Mexico. In Hough, J. L. and H. W. Menard, Eds. *Finding Ancient Shorelines,* Soc. Econ. Paleontologists and Mineralogists, Spec. Publ. 3, pp. 2–19.

Guilcher, A., 1969. Pleistocene and Holocene sea level changes, *Earth Science Reviews,* v. 5, pp. 69–97.

Hamilton, Warren, 1969. Mesozoic California and the underflow of Pacific mantle, *Geol. Soc. America Bull.,* v. 80, pp. 2409–2430.

Hedberg, H., 1964. Geologic aspects of origin of petroleum, *Am. Assoc. Petroleum Geologists Bull.,* v. 48, pp. 1755–1803.

————, 1970. Continental margins from viewpoint of the petroleum geologist, *Am. Assoc. Petroleum Geologists Bull.,* v. 54, pp. 3–43.

Heezen, B. C., and C. L. Drake, 1964. Grand Banks slump, *Am. Assoc. Petroleum Geologists Bull.,* v. 48, pp. 221–233.

Heezen, B. C., C. D. Hollister, and W. F. Ruddiman, 1966. Shaping of the continental rise by deep geostrophic contour currents, *Science,* v. 152, pp. 502–508.

Heezen, B. C., and Marie Tharp, 1965. Physiographic diagram of the Indian Ocean, the Red Sea, the South China Sea, the Sulu Sea, and the Celebes Sea: Geol. Soc. America.

Heezen, B. C., and Marie Tharp, 1968. Physiographic diagram of the North Atlantic Ocean: Geol. Soc. America.

Hoskins, E. G., and J. R. Griffiths, 1971. Hydrocarbon potential of northern and central California offshore. In Cram, I. H., Ed., *Future Petroleum Provinces of the United States—Their Geology and Potential,* Am. Assoc. Petroleum Geologists Memoir 15, v. 1, pp. 212–228.

Hsü, K. J., 1968. Principles of mélanges and their bearing on the Franciscan-Knoxville paradox, *Geol. Soc. American Bull.,* v. 79, pp. 1063–1074.

————, 1971. Origin of the Alps and western Mediterranean, *Nature,* v. 233, pp. 44–48.

Jelgersma, S., 1966. Sea-level changes during the last 10,000 years: Proceedings Internat. Symp. "World climate from 8000 to 0 BC," Imperial college, London, 18–19 April, 1966, Royal Meteorol. Soc. London, pp. 54–71.

Karig, D. E., 1970. Ridges and basins of the Tonga–Kermadec island arc system, *Jour. Geophys. Research,* v. 75, pp. 239–254.

Kay, M., 1951. North American geosynclines, *Geol. Soc. America Memoir* 48, 143 pp.

Krynine, P. D., 1951. A critique of geotectonic elements, *Trans. American Geophys. Union,* v. 32, pp. 743–748.

Lehner, P., 1969. Salt tectonics and Pleistocene stratigraphy on continental slope of northern Gulf of Mexico, *Am. Assoc. Petroleum Geologists Bull.,* v. 53, pp. 2431–2479.

Leopold, L. B., M. G. Wolman, and J. P. Miller, 1964. *Fluvial processes in Geomorphology,* San Francisco, W. H. Freeman and Co., 522 pp.

Luyendyk, B. P., 1970. Origin and history of abyssal hills in the northeast Pacific Ocean, *Geol. Soc. America Bull.*, v. 81, pp. 2237–2260.

Mackin, J. H., 1948. Concept of the graded river, *Geol. Soc. America Bull.*, v. 59, pp. 463–512.

Maloney, N. J., 1967. Continental borderland of Venezuela, Abst., *Geol. Soc. America Special Paper 101*, p. 131.

McFarlan, E., Jr., 1961. Radiocarbon dating of Late Quaternary deposits, south Louisiana, *Geol. Soc. America Bull.*, v. 72, pp. 129–158.

McKenzie, D. P., and W. J. Morgan, 1969. Evolution of triple junctions, *Nature*, v. 224, pp. 125–133.

Meade, R. H., 1969. Landward transport of bottom sediments in estuaries, of the Atlantic coastal plain, *Jour. Sed. Petrology*, v. 39, pp. 222–234.

Menard, H. W., 1964. *Marine Geology of the Pacific*, New York, McGraw-Hill, 271 pp.

Milliman, J. D., and K. O. Emery, 1968. Sea levels during the past 35,000 years, *Science*, v. 162, pp. 1121–1123.

Mitchell, A. H., and H. G. Reading, 1969. Continental margins, geosynclines, and ocean floor spreading, *Jour. Geology*, v. 77, pp. 629–646.

Mitchell, A. H., and H. G. Reading, 1971. Evolution of island arcs, *Jour. Geology*, v. 79, pp. 253–284.

Moore, D. G., 1961. Submarine slumps, *Jour. Sed. Petrology*, v. 31, pp. 343–357.

———, 1969. Reflection profiling studies of the California continental borderland: structure and Quaternary turbidite basins, *Geol. Soc. America Spec. Paper 107*, 142 pp.

———, 1973. Plate edge deformation and crustal growth, Gulf of California structural province, *Geol. Soc. America Bull.*, v. 84, pp. 1883–1906.

Moore, D. G., and J. R. Curray, 1963. Structural framework of continental terrace, northwest Gulf of Mexico, *Jour. Geophys. Research*, v. 68, pp. 1725–1747.

Moore, T. C., Tj. H. van Andel, W. H. Blow, and G. R. Heath, 1970. Large submarine slide off northeastern continental margin of Brazil, *Am. Assoc. Petroleum Geologists Bull.*, v. 54, pp. 125–128.

Mörner, N. A., 1969. Climatic and eustatic changes during the last 15,000 years, *Geol. en Mijnbouw*, v. 48, pp. 389–399.

———, 1971. The Holocene eustatic sea level problem, *Geol. en Mijnbouw*, v. 50, pp. 699–702.

Neumann, A. C., 1969. Quaternary sea-level data from Bermuda, *Abstracts, INQUA VIII Congress*, Paris, 1969, pp. 228–229.

Normark, W. R., and D. J. W. Piper, 1972. Sediments and growth pattern of Navy Deep-Sea Fan, San Clemente Basin, California Borderland, *Jour. Geology*, v. 80, pp. 198–223.

Page, B. M., 1972. Oceanic crust and mantle fragment in subduction complex near San Luis Obispo, California, *Geol. Soc. America Bull.*, v. 83, pp. 957–972.

Phillips, R. P., 1964. Seismic refraction studies in Gulf of California. In Tj. H. van Andel, and G. G. Shor, Jr., Eds., *Marine Geology of the Gulf of California—A symposium,* Tulsa, Am. Assoc. Petroleum Geologists Memoir 3, pp. 90–121.

Postma, H., 1967. Sediment transport and sedimentation in the estuarine environment. In G. H. Lauff, Ed., *Estuaries,* Amer. Assoc. for the Advancement of Science, Pub. 83, pp. 158–179.

Schofield, J. C., 1964. Post-glacial sea levels and isostatic uplift, *New Zealand Jour. Geol. Geophys.,* v. 7, pp. 359–370.

Scholl, D. W., F. C. Craighead, and M. Stuiver, 1969. Florida submergence curve revised: its relation to coastal sedimentation rates, *Science,* v. 163, pp. 562–564.

Schubel, J. K., 1971. The estuarine environment—estuaries and estuarine sedimentation, Short course lecture notes, 30–31 October, 1971, American Geological Institute, Washington, D.C.

Shepard, F. P., 1932. Sediments of the continental shelves, *Geol. Soc. America Bull.,* v. 43, pp. 1017–1040.

———, 1963a. *Submarine Geology,* New York, Harper and Row, 557 pp.

———, 1963b. Thirty-five thousand years of sea level. In T. Clements, Ed., *Essays in Marine Geology in Honor of K. O. Emery,* Los Angeles, Univ. S. Calif. Press, pp. 1–10.

Shepard, F. P., and J. R. Curray, 1967. Carbon-14 determination of sea level changes in stable areas. In *Progress in Oceanography,* v. 4, *The Quaternary History of the Ocean Basins,* Oxford, Pergamon Press, pp. 283–291.

Shepard, F. P., and R. F. Dill, 1966. *Submarine Canyons and Other Sea Valleys,* Chicago, Rand McNally and Co., 381 pp.

Shepard, F. P., R. F. Dill, and B. C. Heezen, 1968. Diapiric intrusions in foreset slope sediments off Magdalena delta, Colombia, *Am. Assoc. Petroleum Geologists Bull.,* v. 52, pp. 2197–2207.

Shepard, F. P., R. F. Dill, and U. von Rad, 1969. Physiography and sedimentary processes of La Jolla submarine fan and fan-valley, California, *Am. Assoc. Petroleum Geologists Bull.,* v. 53, pp. 390–426.

Silver, E. A., 1971. Transitional tectonics and late Cenozoic structure of the continental margin off northernmost California, *Geol. Soc. America Bull.,* v. 82, pp. 1–22.

Silver, E. A., J. R. Curray, and A. K. Cooper, 1971. Tectonic development of the continental margin off central California. In J. H. Lipps and E. M. Moores, Ed., *Geologic Guide to the Northern Coast Ranges, Point Reyes Region, California,* Geological Soc. of Sacramento, pp. 1–10.

Stanley, D. J., 1969. Submarine channel deposits and their fossil analogs ("fluxoturbidites"). In D. J. Stanley, Ed., *The New Concepts of Continental Margin Sedimentation,* American Geological Institute, Washington, D.C., lecture no. 9.

Stanley, D. J., H. Sheng, and C. P. Pedraza, 1971. Lower continental rise east of the middle Atlantic states: predominant sediment dispersal perpendicular to isobaths, *Geol. Soc. America Bull.,* v. 82, pp. 1831–1840.

Stride, A. H., 1963. Current swept sea floors near the southern Gulf of Great Britain, *Quart. Jour. Geol. Soc., London,* v. 119, pp. 175–199.

Suggate, R. P., 1968. Post-glacial sea-level rise in the Christchurch Metropolitan area, New Zealand, *Geol. en Mijnbouw,* v. 47, pp. 291–297.

Swift, D. J. P., D. J. Stanley, and J. R. Curray, 1971. Relict sediments on continental shelves: a reconsideration, *Jour. of Geology,* v. 79, pp. 322–346.

Uchupi, E., 1967a. Slumping on the continental margin southeast of Long Island, New York, *Deep-Sea Research,* v. 14, pp. 635–639.

———, 1967b. The continental margin south of Cape Hatteras, North Carolina: shallow structure, *Southeastern Geology,* v. 8, pp. 155–171.

Uchupi, E., and K. O. Emery, 1968. Structure of continental margin off Gulf Coast of United States, *Am. Assoc. Petroleum Geologists Bull.,* v. 52, pp. 1162–1193.

Van Andel, Tj. H., and J. J. Veevers, 1967. Morphology and sediments of the Timor Sea: Dept. Nat. Development, Bureau Mineral Resources, *Geology and Geophysics Bull.,* no. 83, 173 pp.

Ward, W. T., 1971. Postglacial changes in level of land and sea, *Geol. en Mijnbouw,* v. 50, pp. 703–718.

Wilson, J. T., 1968. Static or mobile earth: the current scientific revolution, *Proc. of the American Philosophical Soc.,* v. 112, pp. 309–320.

While the sediments as a whole are important to petroleum—as source rocks, avenues of migration, reservoirs, and sealing caps over reservoirs—special interest attaches to the organic fraction, from which oil and gas are derived. J. Gordon Erdman, long involved in fundamental chemical studies of petroleum generation, reviews the general processes by which the small organic fraction of the sediment is slowly and partially transformed into hydrocarbons, by inorganic processes, depending on temperature, degree of oxidation, and the mineral matrix. Erdman suggests that the important phase of petroleum generation occurs after the main episode of dewatering.

Geochemical Formation
of Oil

J. Gordon Erdman[1]

INTRODUCTION

Two concepts that have evolved from the work of petroleum geologists and chemists—geochemists as they are called today—are:

1. Petroleum has its origin in source rocks, fine-grained rocks with well defined limits in composition and properties.

2. Genesis involves abiogenic chemical reactions which produce migratable oil only after compaction of the source rock is well advanced.

In short, true petroleum, at least liquid petroleum, is not indigenous to recent uncompacted sediments and is not available for migration until the rocks have become buried and compacted.

For conciseness, several definitions of terms to be used in this paper are provided:

Source rocks. Rocks in which significant quantities of petroleum have been generated and from which the petroleum has migrated.

Biogenic. Produced by the action of organisms.

[1] Phillips Petroleum Corporation, Bartlesville, Oklahoma.

Organic. Compounds characterized by carbon to hydrogen bonds. There is no life connotation.

Hydrocarbons. Compounds consisting solely of the elements carbon and hydrogen. Hydrocarbons make up part but not all of petroleum.

Kerobitumens. Dark brown to black compounds of high molecular weight occurring as a residue and by-product of petroleum genesis.

The objective of this paper is to examine some of the basic geochemistry of petroleum genesis to establish (1) boundaries for hypotheses and (2) to suggest areas of needed research.

Petroleum geochemistry has come of age in that the basic concepts currently held provide the foundation for a variety of successful exploration methods. Particularly gratifying is that whereas many of these methods are proprietary, comparison of results and conclusions, as for example at the symposium on the characterization of the cores from the Challenger Knoll in the Gulf of Mexico, have shown excellent agreement (Davis and Bray, 1968).

RATE AND EXTENT OF GENESIS

An analysis of the known variables that determine the extent and rate of petroleum genesis are shown in Tables I and II. All of these variables have been recognized and considered by geologists and geochemists for many years, but an overview to determine the specific effects of each in relation to the others only now is being developed. The five fundamental variables listed in Table II are discussed below.

Concentration of Organic Matter in the Source Rock

Rocks of marine origin generally do not contain more than a few percent indigenous organic matter. True marine counterparts of the lignites and coals of terrestrial origin, consisting predominantly of organic matter, are very rare and are not a factor in the formation of petroleum accumulations.

For reasons to be discussed later, petroleum genesis at best is an inefficient process, that is, the yield is low in terms of the organic matter initially incorporated into the rock. Most of this organic matter survives and is retained in the source rock as a brown to black amorphous material that we call kerobitumen. Thus the organic richness of a rock probably does not decrease to any large extent with genesis and migration of petroleum.

TABLE I.
Parameters controlling genesis of petroleum.

YIELD OF OIL   IS A FUNCTION OF: (1) THE CONCENTRATION OF ORGANIC
(SOURCE ROCK)   MATTER IN THE SOURCE ROCK

                                               (2) THE RATE OF CONVERSION OF THE
                                                   ORGANIC MATTER TO OIL

                                               (3) GEOLOGIC TIME INTERVAL

RATE OF CONVERSION   IS A FUNCTION OF: (1) TEMPERATURE
OF THE ORGANIC MATTER
TO OIL   (2) SUSCEPTIBILITY TO CONVERSION
                                                   TO PETROLEUM

                                               (3) MINERAL MATRIX

SUSCEPTIBILITY TO   IS A FUNCTION OF: (1) COMPOSITION OF THE PRODUCTS
CONVERSION TO PETROLEUM   OF BIOSYNTHESIS (ESSENTIALLY
                                                   A CONSTANT FOR MARINE AND
                                                   LITTORAL ENVIRONMENTS)

                                               (2) EXTENT OF ABIOGENIC OXIDATION
                                                   BEFORE, DURING, AND IMMEDIATELY
                                                   FOLLOWING DEPOSITION

TABLE II
Parameters controlling genesis of petroleum.

YIELD OF OIL   IS A FUNCTION OF:
SOURCE ROCK

                      (1) THE CONCENTRATION OF ORGANIC
                           MATTER IN THE SOURCE ROCK

                      (2) THE GEOLOGIC TIME INTERVAL

                      (3) TEMPERATURE

                      (4) EXTENT OF ABIOGENIC OXIDATION
                           BEFORE, DURING, AND IMMEDIATELY
                           FOLLOWING DEPOSITION

                      (5) MINERAL MATRIX

To be a source rock, namely a rock that has yielded oil, hopefully to an adjacent reservoir, sufficient petroleum must be generated to migrate. The interval in geologic time and the mechanism by which this migration occurs has been the center of much debate for several decades and now is under critical study in many laboratories. At the present time, it can be said with certainty that: (1) petroleum genesis is a slow process and that sufficient liquid petroleum normally is not generated until after the rock is deeply buried and compaction is well advanced; and (2) only a part of the liquid petroleum generated migrates, the proportion depending upon the physical and chemical properties of the rock and the pressure gradients imposed upon it. In all probability most of the petroleum existing in reservoirs has been generated in rocks which contained less than 1.5% organic matter.

Geologic Time Interval—Temperature

These two variables are so closely interrelated that they are best discussed together. For any given reaction, temperature can be traded for time. The terms of the trade are determined by the activation energy. An empirical rule among organic chemists is that an increase in temperature of about 10°C (18°F) will double the rate of reaction. This empirical rule has been carried over into geochemistry and used to construct curves such as the one shown in Figure 1 relating the geologic time necessary for petroleum genesis to depth of burial or more precisely temperature.

The particular curve in Figure 1 was constructed on the basis of an observation that in a Lower Cretaceous rock at a present depth of 1500 m (4920 ft) and a geothermal gradient of 2°C/100 m (1.2°F/100 feet) petroleum has been generated in sufficient quantity to provide a commercial accumulation in an adjacent trap. Following along the curve, we can see that for Middle Miocene rocks to achieve comparable genesis the depth would have to be about 2600 m (8520 ft) whereas for Middle Ordovician rocks genesis would have been accomplished at only 530 m (1740 ft). For the same degree of genesis to take place in the one million years to the beginning of the Pleistocene would require a depth of burial of 5400 m (16,400 ft) and a present formation temperature of 124°C (255°F). The geological time scale is after Kulp (1961).

Relations of this sort involve four assumptions. First, it must be assumed that the chemical composition of the organic fraction and the mineral composition of the sediment matrix are constant, that is, the same in the Middle Miocene as in the Lower Cretaceous, as in the Middle Ordovician. Second, it must be assumed that the depositional histories are the same, except compressed in respect to depth and time. Third, it must be assumed that the geothermal gradient and surface temperature have remained constant. Finally, it must be assumed that the empirical rule of 10°C (18°F) to double the rate actually represents

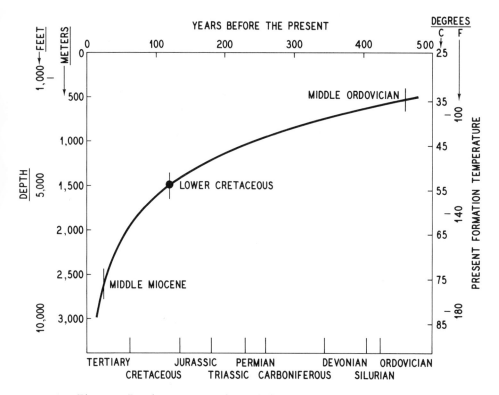

Figure 1. Depth-temperature-time relationship for petroleum genesis.

the average for the geochemical reactions taking place and that no new reactions with higher activation energies enter into the diagenesis with increasing depth of burial and temperature.

Seldom are the above conditions realized in nature. Expanding on the last condition, the rapid burial with concomitant rapid increase in temperature necessary to effect genesis in a short time in accord with Figure 1 could initiate and advance reactions destructive to petroleum at rates approaching or equal to the rate of genesis. To be specific, in Figure 2 is shown a disproportionation-cyclization reaction whereby an isoprenoid pigment is converted to aromatic hydrocarbons that occur in petroleum (Mulik and Erdman, 1963; Day and Erdman, 1963; Erdman, 1965). This reaction is known to have a low activation energy, that

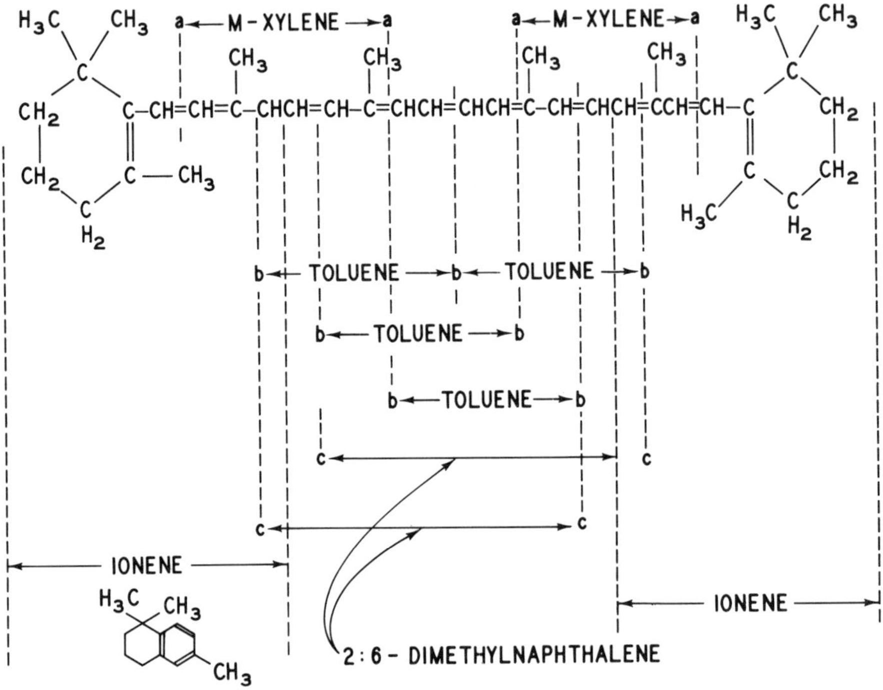

Figure 2. Mechanism of degradation of β-carotene to form aromatic hydrocarbons present in most crude oils.

is, the reaction increases rather slowly with rising temperature. Figure 3, on the other hand, shows reactions with higher activation energies that become important in sedimentary rocks only at intermediate and high temperatures. These reactions would work against petroleum genesis, the upper to block genesis by exhausting source material and the second by destroying any petroleum already generated. Thus, for each sediment composition, there must be a minimum time for genesis to take place.

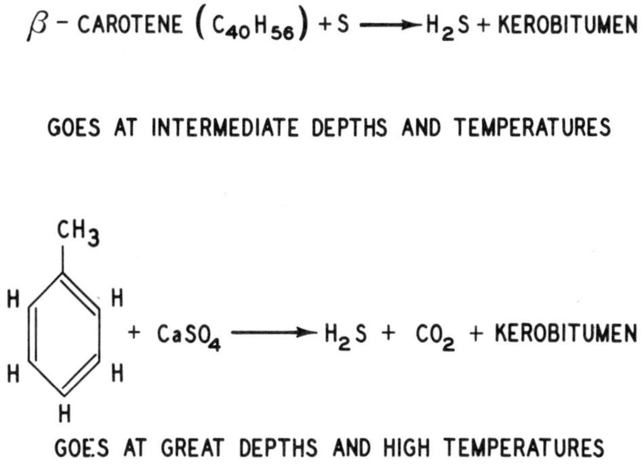
β - CAROTENE (C40H56) + S ⟶ H2S + KEROBITUMEN
GOES AT INTERMEDIATE DEPTHS AND TEMPERATURES
CH3
H
H
H
H
H
+ CaSO4 ⟶ H2S + CO2 + KEROBITUMEN
GOES AT GREAT DEPTHS AND HIGH TEMPERATURES

Figure 3. Reactions that work against petroleum genesis.

Extent of Abiogenic Oxidation Before, During and Immediately Following Deposition

In Figures 4 and 5 are plotted average Odd-Even Predominance, $\overline{\text{OEP}}$, values versus formation temperatures for rock samples from two Tertiary basins. Average Odd-Even Predominance, developed in the author's laboratory, is a mathematically correct statement, provided in the form of a single number, of the predominance in the n-alkane series of odd over even carbon numbers (Scalan and Smith, 1969). Like the Correlation Preference Index (Bray and Evans, 1961; Cooper and Bray, 1963), unity represents no predominance; numbers greater than unity, odd carbon number predominance; and numbers less than unity, even carbon number predominance. Normally, for fine-grained rocks, $\overline{\text{OEP}}$ and CPI values decrease toward unity with increasing age and depth of burial. During the early stages of diagenesis, such decrease is considered suggestive of dilution with abiogenically generated oil having a value of unity.

Returning to Figures 4 and 5, we note that both sets of samples are from thick clastic sections of approximately the same geologic age, geothermal gradient, and organic richness. In Figure 4 the value of $\overline{\text{OEP}}$ tends to decrease toward unity but the rate of decline is slow and even at the highest measured temperatures, that is, maximum sampling

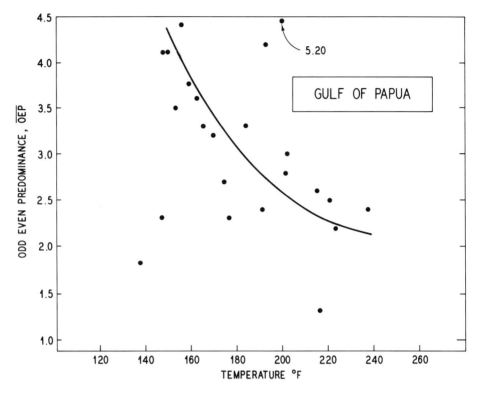

Figure 4. Papuan basin Tertiary clastics. Odd-even predominance versus formation temperature.

depths, still shows relatively high values. It perhaps is not surprising that little or no liquid petroleum has been found associated with the Tertiary in this basin. In Figure 5 the $\overline{\text{OEP}}$ values decrease rapidly toward unity with increasing temperature. The lower portion of the Tertiary in this basin is a prolific source of oil.

These findings indicate that caution is in order in evaluating the prospects of a given area on the basis of a depth ceiling or temperature floor derived from averaging the temperatures of reservoirs over a large area such as was done by Burst (1969) and utilization of such data in another, albeit adjacent area as was done by Reel and Griffin (1971) in their recent article on the prospects of Florida.

But why is petroleum genesis proceeding at relatively low temperatures in one basin, whereas in another basin in apparently the same

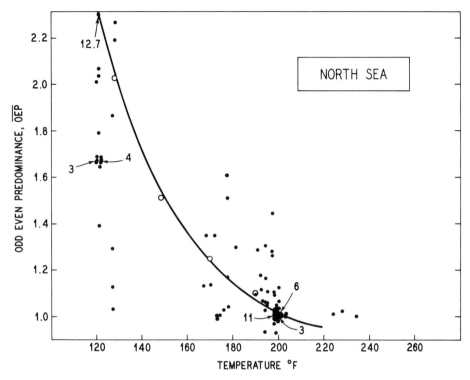

Figure 5. North Sea basin Tertiary clastics.
Odd-even predominance versus formation temperature.

kind of rocks and at even higher temperatures it is proceeding much more slowly if at all?

First of all there is a myth that has become widespread in discussions of the genesis of petroleum, namely that marine microorganisms are capable of synthesizing many if not all of the compounds that constitute petroleum. The function of the enzyme systems of organisms is to produce energy and to synthesize more of their life substance for growth and reproduction, primarily lipids, proteins, and carbohydrates. In aerobic environments, utilization of each of these three major classes of natural products is in proportion to their abundance. In the ocean there may be an attrition of 99.9% between the euphotic zone and the bottom, yet there is little difference in the chemical composition of euphotic and benthonic organisms. In environments of restricted oxygen supply, there is a tendency to preferential metabolism of oxygen-rich compounds such as carbohydrates with concentration of lipids. In sediments, both aerobic and anerobic life processes soon are terminated by the biostatic action of older metabolic products to which organisms are exposed during the early stages of compaction. This mechanism is shown schematically in Figure 6 (Erdman, 1965). Basically, the abiogenic phase of organic diagenesis in the earth begins with the three major classes of natural products: lipids, proteins, and carbohydrates.

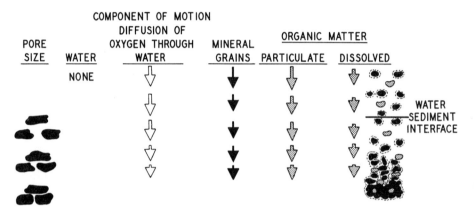

Figure 6. Relative motion of components of an aquatic sediment during compaction. (Condition of uniform sedimentation in a stable environment.)

An understanding of the boundary conditions for petroleum genesis, or what might be called a "chemical fence," can be achieved from a consideration of the stoichiometry. At the left of Figure 7 are shown the formulas for stearic acid, a typical lipid component; alanine, an amino acid building block of protein; and glucose, a carbohydrate. As a representative constituent of petroleum, methane is used here. Shown on the right are the by-products, water or carbon dioxide, ammonia in the case of the alanine, and kerobitumen. The ratio of the carbon, hydrogen, and oxygen in the kerobitumen is taken from an analysis, shown in Table III, of the kerobitumen of the Cretaceous La Luna limestone of Venezuela at a location where it is considered to be the source rock of the La Luna crude. In each case, the equations have been balanced and the yields of methane calculated.

For simplicity, in Figure 7 the nitrogen and sulfur content of the kero-

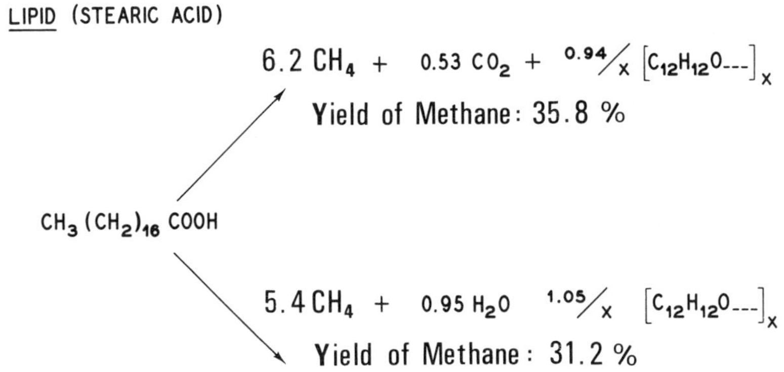

LIPID (STEARIC ACID)

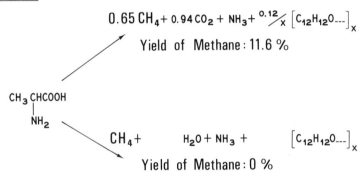

$$6.2\ CH_4\ +\quad 0.53\ CO_2\ +\ ^{0.94}\!\!/_x\ \left[C_{12}H_{12}O\text{---}\right]_x$$

Yield of Methane: 35.8 %

$$CH_3(CH_2)_{16}\ COOH$$

$$5.4\ CH_4\ +\quad 0.95\ H_2O\quad ^{1.05}\!\!/_x\ \left[C_{12}H_{12}O\text{---}\right]_x$$

Yield of Methane : 31.2 %

Figure 7A. For the lipid, stearic acid, maximum and minimum yields (assuming 100% efficiency) of the petroleum hydrocarbon, methane.

α-AMINO ACID (ALANINE)

$$0.65\ CH_4 + 0.94\ CO_2 + NH_3 + ^{0.12}\!\!/_x\ \left[C_{12}H_{12}O\text{---}\right]_x$$

Yield of Methane : 11.6 %

$$CH_3\ CHCOOH$$
$$\quad\quad |$$
$$\quad\quad NH_2$$

$$CH_4 +\quad\quad H_2O + NH_3 +\quad\quad \left[C_{12}H_{12}O\text{---}\right]_x$$

Yield of Methane : 0 %

Figure 7B. For the α-amino acid, alanine, maximum and minimum yields (assuming 100% efficiency) of the petroleum hydrocarbon, methane.

CARBOHYDRATE (GLUCOSE)

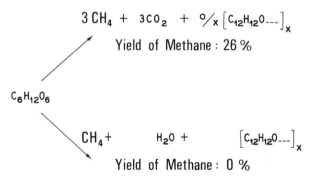

$$3\ CH_4\ +\quad 3\ CO_2\ +\ ^o\!\!/_x\ \left[C_{12}H_{12}O\text{---}\right]_x$$

Yield of Methane : 26 %

$$C_6H_{12}O_6$$

$$CH_4 +\quad\quad H_2O +\quad\quad \left[C_{12}H_{12}O\text{---}\right]_x$$

Yield of Methane : 0 %

Figure 7C. For the carbohydrate, glucose, maximum and minimum yields (assuming 100% efficiency) of the petroleum hydrocarbon methane.

bitumen was not taken into consideration. The nitrogen can be provided from the ammonia generated in the amino acid (protein) degradation. Lipids, proteins, and carbohydrates, however, do not contain sufficient sulfur to provide the sulfur that occurs in petroleum and its kerobitumen by-product. Most of this sulfur must be provided by biogenic reduction of sulfate to sulfur via hydrogen sulfide followed by abiogenic reaction of the elemental sulfur with the organic matter. In the following discussion and calculations the full formula for the kerobitumen, namely $[C_{12}H_{12}ON_{0.16}S_{0.43}]_x$ has been used.

TABLE III
Elemental composition of the insoluble
organic fraction of LaLuna shale
(kerobitumen).

ELEMENTAL ANALYSIS:

CARBON _____ 78.2 %

HYDROGEN _____ 6.2 %

OXYGEN _____ 7.6 %

NITROGEN _____ 1.13 %

SULFUR _____ 6.89 %

APPROXIMATE FORMULA $C_{12} H_{12} O N_{0.16} S_{0.43}$

In Figure 8, the yields of open chain alkane from methane, carbon number 1, to heptadecane, carbon number 17, are plotted for the degradation of the lipid, stearic acid. The upper edge of the stippled band represents the reaction involving elimination of oxygen as carbon dioxide; and the lower edge of the band by elimination of oxygen as water. The area within the band represents mixed yields of carbon dioxide and

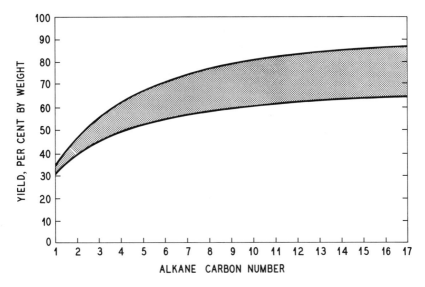

Figure 8. Degradation of stearic acid. Yield of hydrocarbon.

water. In these reactions the yield of hydrocarbon is determined by the amount of available hydrogen. Methane, CH_4, has the highest hydrogen to carbon ratio. The ratio decreases progressively with increasing chain length—ethane, C_2H_6; propane, C_3H_8; butane, C_4H_{10}; etc.; hence with increasing molecular weight, the yield increases toward a constant value for $(CH_2)_x$. As the molecular weight of the hydrocarbon and the yield increase, the proportion of kerobitumen must necessarily decrease, reaching zero at C_{17} on the upper edge of the band. Here the reaction represents simply decarboxylation of the acid. Under all other conditions there is formation of kerobitumen.

In geochemical characterization of source rocks the amount of residual oil and kerobitumen is determined and, from these data, the ratio of oil to total organic matter. Information desired, however, is the amount of additional oil generated and lost by migration. An objective of this discussion is to establish limiting ranges. In Figure 9, the data from the equations used to derive Figure 8 have been re-computed as a range of ratios of hydrocarbon to total residual organic matter, which is the sum of the hydrocarbon and the kerobitumen. The hatched band encompasses the ratios possible between the extreme case where the

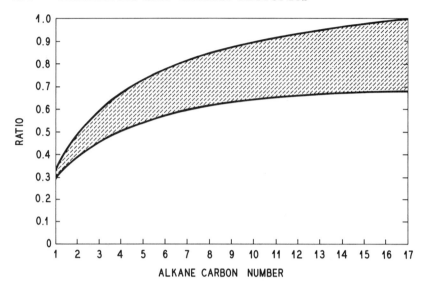

Figure 9. Degradation of stearic acid. Ratio of hydrocarbon to total organic matter.

excess oxygen is eliminated as carbon dioxide and where it is eliminated as water. The value of unity at C_{17} for the carbon dioxide elimination reaction again represents the unique case where the hydrocarbon itself is the only organic product.

The stoichiometry for the degradation of the amino acid, alanine, to form the same series of hydrocarbons gives quite different patterns. Figure 10 represents the same type of plot as Figure 8. In this case the lower boundary of the yield band is zero throughout the entire range. For the reaction proceeding by elimination of oxygen as water, there is not even enough hydrogen to generate the kerobitumen. Either a kerobitumen of lower hydrogen-higher oxygen content must be formed or the reaction will be parasitic in regard to hydrogen, that is, it will derive hydrogen from a hydrogen richer co-natural product such as stearic acid. In so doing, it decreases the capacity of that compound to generate hydrocarbons. The extent of this decrease is depicted by the checked band. The upper limit of values for the ratio of hydrocarbon to total organic matter is shown in Figure 11.

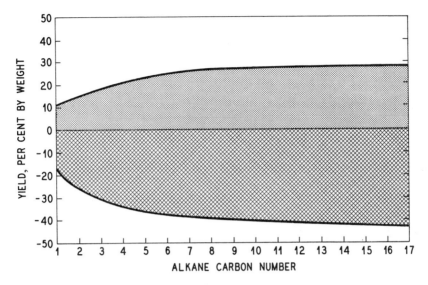

Figure 10. Degradation of alanine. Yield of hydrocarbon.

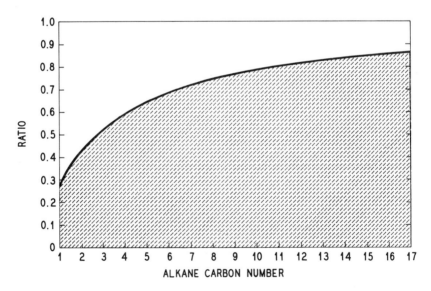

Figure 11. Degradation of alanine. Ratio of hydrocarbon to total organic matter.

The degradation of the carbohydrate, glucose ($C_6H_{12}O_6$), provides a still more complex pattern as shown in Figure 12. Glucose is rich in oxygen and half the carbon in the molecule is required to eliminate it as carbon dioxide. After doing so, there remains sufficient hydrogen to convert the remaining carbon to methane and an excess of hydrogen for forming the higher hydrocarbons. The yield of hydrocarbon as depicted by the upper limit of the band is nearly constant from methane

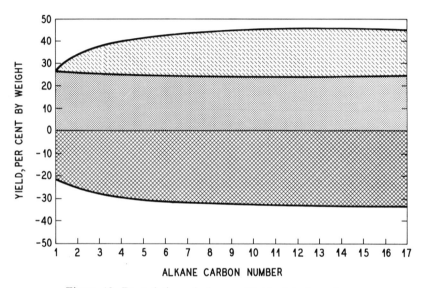

Figure 12. Degradation of glucose. Yield of hydrocarbon.

through heptadecane. No kerobitumen is formed, and the excess hydrogen would be donated to another degradation reaction such as that of alanine. The reaction whereby excess oxygen is eliminated as water resembles the reaction for alanine in that there is not even enough hydrogen to form the kerobitumen. Again either a kerobitumen of lower hydrogen-higher oxygen content must be formed or the reaction will be parasitic in regard to hydrogen on other natural product precursors, such as stearic acid. In regard to the ratio of hydrocarbon to total organic product (hydrocarbon plus kerobitumen) the limits are unity and zero throughout the entire range.

Thus far in this discussion it has been assumed that no abiogenic oxidation of the source material during and immediately following deposition has occurred. If we refer to the sedimentation mechanism, Figure 6, it can be seen that oxidation to a varying degree might well take place. The highest probability for oxidation will be in organic-rich sediments with a spongy texture. A peat bog constitutes an extreme example. In such a bog there is a slow rate of compaction, high concentrations of biostatic agents, and high porosity and permeability favoring rapid diffusion of oxygen.

The calculations have been repeated for stearic acid, alanine, and glucose, introducing increments of molecular oxygen into the reactions. The effect on yield of hydrocarbon and kerobitumen vary. The overall effect is a decrease in yield of hydrocarbon and an increase in kerobitumen, which by necessity must become progressively more depleted in hydrogen and/or enriched in oxygen.

The nonhydrocarbon asphaltic portion of crude oils, consisting of resins and asphaltenes, compositionally resembles the source rock kerobitumen except that the hydrogen to carbon ratio usually is higher and the oxygen content lower. In the reactions discussed, formation of resins and asphaltenes occurs at the expense of hydrocarbon and kerobitumen. The shape of the yield curves in Figures 8, 10, and 12, however, would not change. As shown in Table IV the petroleum resins, petroleum asphaltenes, source rock kerobitumen, and coals tend to form a series. To return to petroleum genesis, abiogenic oxidation at the time of or shortly following deposition would work against the formation of low molecular weight hydrocarbons and would lead to formation of heavy crudes. Not only would the limits of petroleum genesis be decreased but the rate of genesis of petroleum including n-paraffins, upon which OEP is based, would appear to take place more slowly. According to this concept, the organic matter deposited in the Papuan basin was subjected to considerable oxidation. In the North Sea Tertiary basin, on the other hand, there probably was much less oxidation and the source material was unusually high in lipids.

The type of analysis of source and yield of petroleum discussed here is limited by the uncertainty concerning the proportion of the oxygen in oxygen-containing precursors, or source compounds, that is degraded to carbon dioxide and to water. In the case of lipids where the oxygen content of the precursor is low, the effect on the yield of hydrocarbon and the ratio of oil to total organic matter are not overly large. For protein and carbohydrates, the effect is so large as to control whether these families of natural products will contribute to petroleum formation or prove a detriment.

Carbon isotopic data for crude oils and the lipid and total carbon of marine plants are shown in Figure 13. These data suggest that the lipids may be the major precursor and in turn that the principal route

TABLE IV

Composition of petroleum asphaltics, source rock kerobitumens, lignite, and coals.

SAMPLE	ATOM RATIOS			
	H/C	O/C	N/C	S/C
PETROLEUM RESIN				
BAXTERVILLE	1.27	0.0044	0.0051	0.022
PETROLEUM ASPHALTENE				
BAXTERVILLE	1.05	0.015	0.0081	0.025
MARA	1.13	0.0089	0.0093	0.024
LAGUNILLAS	1.13	0.014	0.020	0.020
WAFRA NO. 17	1.18	0.012	0.0057	0.036
BURGAN	1.17	0.0056	0.018	0.035
RAUDHATAIN	1.18	0.016	0.0087	0.036
RAGUSA	1.29	0.016	0.015	0.029
LIMESTONE - MARINE				
LA LUNA	0.95	0.074	0.012	0.033
LIGNITE				
KINCAID MINE	0.95	0.26	0.0076	0.00022
BITUMINOUS COALS				
LOWER FREEPORT	0.99	0.079	0.0067	0.0018
UPPER FREEPORT	0.77	0.064	0.0018	0.0030
HANNA	1.11	0.20	0.015	0.0033

of degradation involves elimination of excess oxygen as carbon dioxide. Model reactions worked out in the laboratory, such as that shown in Figure 14 (Erdman, 1967), also suggest that the route may be via carbon dioxide, but they are not conclusive because degradation in the earth almost certainly comprises a complex of reactions. The best answer to the question would be determination of the relative free energies, ΔF, for the reactions.

Because the structures of petroleum resins, asphaltenes, and rock kerobitumens include complex aromatic ring systems, their free energies cannot be computed from bond energies. Determination of heats of combustion, ΔH, and heat capacities are required. These compounds are difficult to burn cleanly to simple products, hence combustion calorimetry would be difficult. Perhaps this discussion will stimulate experimental thermodynamists to undertake this formidable task. As an added incentive, establishing the quantity of carbon dioxide generated as by-product

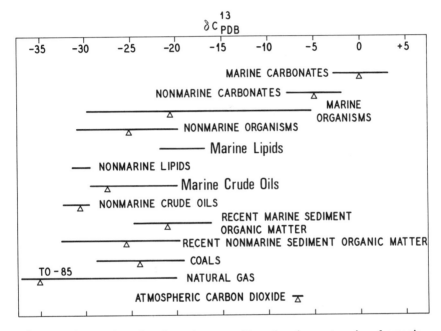

Figure 13. Ranges in carbon isotopic composition of various categories of organic matter occurring in nature.

in petroleum genesis would contribute to the solution of a related geologic problem, namely whether the carbon dioxide occurring with petroleum in sedimentary reservoirs is derived from degradation of organic matter or stems from thermal decomposition of carbonates or outgassing of basement rocks.

Rock Matrix

The fact that petroleum genesis takes place in the presence of a finely didived mineral matrix early led to the suggestion of catalysis as a major factor in genesis. Cracking and isomerization reactions of the type proposed by Brooks (1948, 1949, 1952) are not likely to occur in the earth because water provides an effective barrier to adsorption of hydrocarbons. Investigations of the degradation of polar compounds such as fatty acids over water wet clay minerals by Eisma and Jurg (1957) are not

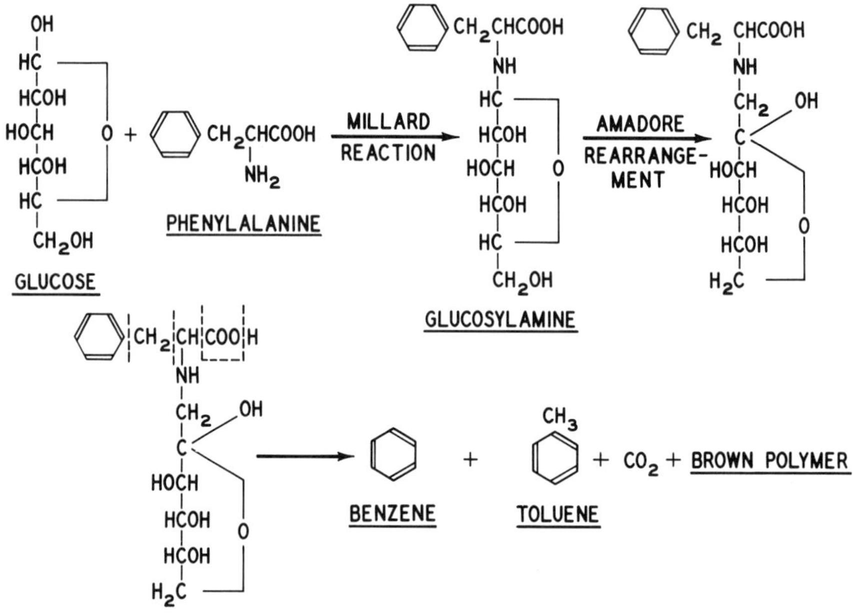

Figure 14. A Mechanism of degradation of glucose (carbohydrate) and phenylalanine (α-amino acid) to form benzene, toluene and a brown polymer that appears to be a precursor of petroleum resins, asphaltenes, and kerobitumen. The reaction appears to be general, that is, all monosaccharides (carbohydrate) and α-amino acids can participate with formation of the entire series of low molecular weight hydrocarbons.

conclusive. Furthermore, petroleum genesis takes place in rocks of widely different mineralogies, shales and mudstones of widely different clay mineral composition, limestones and mud-sized silicas.

Based on the work of Weaver (1960), Burst (1969) and others have assumed that the extent of clay diagenesis can be used as an index of the degree to which genesis of petroleum has taken place provided that sufficient organic matter is present. According to this concept, a rock having a relatively high percentage of mixed-layer illite-montmorillonite in which the ratio of illite to swelling components of the clay is intermediate to high has undergone diagenetic alteration, and consequently the rock has been subjected to deep burial and high temperatures and is likely to be associated with the genesis of oil. Conversely, a rock having a low ratio for the mixed-layer clay would be considered an immature source rock because insufficient diagenesis has occurred. Well crystallized illite would be indicative of metamorphism and destruction of petroleum. This hypothesis is contrary to the work of Grim (1953, 1958), Johns and Grim (1958), and Milne and Early (1958), which suggests that variations in the illite-montmorillonite ratio with depth may be the result of different depositional environments rather than diagenetic alteration. Findings in this laboratory, for example the clay mineral log shown in Figure 15, support the latter view.

Whether or not mineral catalysis is an important factor in petroleum

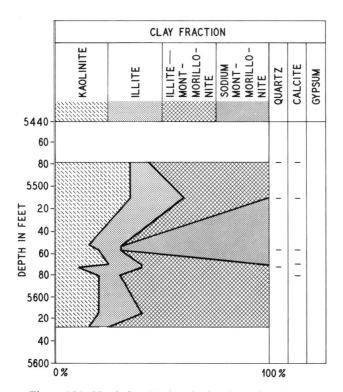

Figure 15A. North Sea Tertiary basin, clay mineral log.

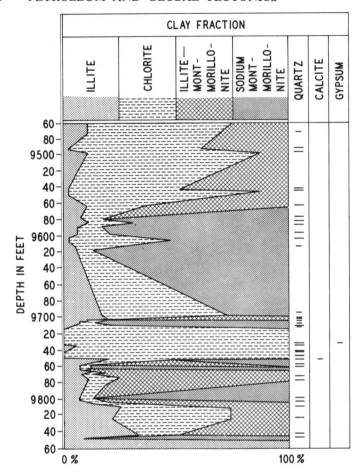

Figure 15B. North Sea Tertiary basin, clay mineral log.

genesis, it is certain that, but for the adsorptive properties of the mineral matrix of source rocks, there would be no petroleum accumulations. Because hydrocarbons and other oily nonpolar compounds are not adsorbed on mineral surfaces, most biologically generated methane and other free hydrocarbons that might be present in newly deposited sediments are swept out, probably in solution, during early compaction. Insoluble, particulate organic matter and polar oganic matter adsorbed on the mineral surfaces are retained. Slow loss of polar groups as part of the genesis process makes the petroleum available when the volume of moving water has been sufficiently reduced to permit migration of petroleum in concentrations favorable for accumulation in reservoirs which also have been in formation and now are ready to receive the petroleum.

CONCLUSION

The concept of global tectonics is a concept of a highly dynamic earth crust, particularly in intercontinent areas. In these areas lie the best remaining prospects for discovery and production of petroleum. Organic richness, the temperature range, extent of oxidation during and following deposition, and mineral matrix constitute the boundaries, the "fence," within which petroleum can be generated and accumulated. Understanding and recognition of this preserve as part of our understanding of global tectonics provides the hope and challenge for fulfilling our petroleum needs for the future.

REFERENCES

Bray, E. E., and E. D. Evans, 1961. Distribution of n-paraffins as a clue to recognition of source beds, *Geochim. Cosmochim. Acta,* v. 22, pp. 2–15.

Brooks, B. T., 1948. The role of surface active catalysis in formation of petroleum I, *Bull Am. Assoc. Pet. Geol.,* v. 32, p. 2269.

———, 1949. The role of surface active catalysis in formation of petroleum II, *Bull. Am. Assoc. Pet. Geol.,* v. 33, p. 1600.

———, 1952. Evidence of catalytic action in petroleum formation, *Ind. Eng. Chem.,* v. 44, p. 2570.

Burst, J. F., 1969. Diagenesis of Gulf Coast clayey sediments and its possible relation to petroleum migration, *Bull. Am. Assoc. Pet. Geol.,* v. 53, pp. 73–93.

Cooper, J. E., and E. E. Bray, 1963. A postulated role of fatty acids in petroleum formation, *Geochim. Cosmochim, Acta,* v. 27, pp. 1113–1127.

Davis, J. B. and E. E. Bray, 1968. Analyses of oil and cap rock from Challenger (Sigsbee) Knoll: Part II, Section 22 in Ewing, M., J. L. Worzel, et al, *Initial Reports of the Deep Sea Drilling Project.* v. I, pp. 415–426.

Day, W. E. and J. G. Erdman, 1963. Ionene: a thermal degradation product of β-carotene, *Science,* v. 141, no. 3583, p. 808.

Eisma, E. and J. W. Jurg, 1957. Fundamental aspects of the diagenesis of organic matter and the formation of hydrocarbons, *Proceedings of the Seventh World Petroleum Congress,* v. 2, pp. 61–77.

Erdman, J. G. and V. G. Ramsey, 1961. Rates of oxidation of petroleum asphaltenes and other bitumens by alkaline permanganate, *Geochim. Cosmochim. Acta*, v. 25, pp. 175–188.

————, and J. P. Dickie, 1964. Mild thermal alteration of asphaltic crude oils, *Preprints of the Petroleum Division, Am. Chem. Soc.*, v. 9, no. 2, pp. B-69.

————, 1965. Petroleum—its origin in the earth. In *Fluids in Subsurface Environments—A Symposium*, Memoir No. 4, Am. Assoc. Pet. Geol.

————, 1967. Geochemical origins of the low molecular weight hydrocarbon constituents of petroleum and natural gases, *Proceedings of the Seventh World Petroleum Congress*, v. 2, pp. 13–24.

Grim, R. E., 1953. *Clay Mineralogy*, N.Y., McGraw-Hill Co., pp. 316–368.

————, 1958. Concept of diagenesis in argillaceous sediments, *Bull. Am. Assoc. Pet. Geol.*, v. 42, pp. 246–253.

Johns, W. D., and R. E. Grim, 1958. Clay mineral composition of Recent sediments from the Mississippi River Delta, *Jour. Sed. Petrology*, v. 28, pp. 180–199.

Kulp, J. L., 1961. Geologic time scale, *Science*, v. 133, p. 1105.

Milne, I. H., and J. W. Earley, 1958. Effect of source and environment on clay minerals, *Bull. Am. Assoc. Pet. Geol.*, v. 42, pp. 328–338.

Mulik, J. D., and J. G. Erdman, 1963. Genesis of hydrocarbons of low molecular weight in organic-rich aquatic systems, *Science*, v. 141, no. 3583, p. 806.

Reel, D. A., and G. M. Griffin, 1971. Potentially petroliferous trends in Florida as defined by geothermal gradients, *Transactions* of the 21st Annual Meeting of the Gulf Coast Association of Geological Societies, New Orleans, October, 1971.

Scalan, R. S., and J. E. Smith, 1970. An improved measure of the odd-even predominance in the normal alkanes of sediment extracts and petroleum, *Geochim. Cosmochim. Acta*, v. 34, pp. 611–620.

Weaver, O. E., 1960. Possible uses of clay minerals in search for oil, *Bull. Am. Assoc. Pet. Geol.*, v. 44, pp. 1505–1518.

Having examined the chemistry of hydrocarbon genesis, including the roles played therein by temperature and time, we return to the tectonic side, to examine the relationships between hydrocarbon occurrence, the tectonic nature of the productive basins, and the thermal regime of these basins. H. Douglas Klemme, with wide experience in petroleum exploration, brings together a large body of data on these subjects. Different tectonic basin types must have different thermal histories, leading to hydrocarbon generation at different stages in their development, and at different depths. He suggests that in general the basins with relatively high heat flow—i.e. the more mobile basins of the continental borders—offer optimal conditions for the generation, migration, and accumulation of petroleum.

Geothermal Gradients, Heat Flow and Hydrocarbon Recovery

H. Douglas Klemme[1]

ABSTRACT

The present and past geothermal gradients in various hydrocarbon-bearing basins appear to have had an influence on the relative magnitude of hydrocarbon recovery. Considerable evidence suggests that high geothermal gradients in clastic rock sequences enhance the processes of formation, migration, and entrapment of oil or gas.

Sedimentary basins may be classified on the basis of plate tectonics. Their geothermal gradients, with exceptions, express the earth's heat flow, the patterns of which are also compatible with the models of plate tectonics. Depth of hydrocarbon occurrence appears to be related to the temperature history of any given basin.

Several basin types associated with significantly high heat flow zones are located along various continental plate margins and areas of possible incipient sea floor spreading or upwelling of basic material. Such basins tend to yield more hydrocarbons per cubic mile of sediments than basins from low heat flow areas if all the proper geologic factors for hydrocarbon accumulation are present.

[1] Lewis G. Weeks Associates, Ltd.

Temperature, modified by time, has been instrumental in the accumulation of many major (giant) accumulations of hydrocarbons.

ACKNOWLEDGMENTS

The writer would like to acknowledge the following sources of data not listed in the bibliography: D. A. Baker, Shell Oil, Denver; R. B. Brown, Broken Hill Proprietary Co. (Oil Division), Melbourne, Australia; J. Cloeter, Arco, New York; W. Del'Oro, Aramco, Dhahran, Saudi Arabia; H. R. Gould, Esso Research, Houston; M. K. Horn, Cities Service Exploration and Production Research, Tulsa; P. Jones, U.S.G.S.—Gulf Coast Hydroscience Center, St. Louis; M. N. Mayuga, Long Beach; E. S. Parker, Chevron Research Laboratory, La Habra; G. T. Philippi, Colorado Springs; K. L. Russel, Exploration Research, Cities Service Oil Co., Tulsa; W. F. B. Ryan, Lamont-Doherty Geological Observatory, Palisades; P. C. Smith, IIAPCO, San Francisco; S. V. Sykes, IOOC, Tehran, Iran; G. H. Taylor, Commonwealth Science and Ind. Research Org., Australia; N. Yn. Uspenskaya, Institute of Petrochemical and Gas Industry, U.S.S.R.; R. Wallace, U.S.G.S.—Gulf Coast Hydroscience Center, St. Louis; and D. Welte, University Göttingen, West Germany. Help in the form of discussions and constructive criticism was provided by A. G. Fischer, H. D. Hedberg, R. E. King, and L. G. Weeks.

INTRODUCTION

This will primarily be a discussion of the relation of temperature to the formation, migration, and accumulation of hydrocarbons. Sedimentary basins represent a variety of tectonic settings and styles, which appear to be related to crustal plate motions, accretion, and destruction. The heat flow in basins, and its partial expression in geothermal gradients, shows a correspondence to these basinal types. A survey of hydrocarbon occurrence relative to basin type and thermal regimes leads to the conclusion that the high geothermal gradients in basins of high heat flow enhance the efficiency of petroleum generation and entrapment, providing the essential geologic conditions are present.

HEAT FLOW IN GENERAL—RELATED TO
SEA FLOOR SPREADING

Heat flow is the transfer of thermal energy from the interior of the earth to the surface, where it is dissipated. It is expressed in the rate of increase of temperature with depth (geothermal gradient) as related to the thermal conductivity of the rocks in which it is measured. Oceans and continents appear to have the same mean flow (approximately 1.5 microcalories per square centimeter per second). Patterns of variation in recorded heat flow measurements include decrease in the heat flow away from oceanic ridges, low heat flow in the vicinity of subduction trenches, high heat flow in the back side of island arcs, and high heat flow in rift zones. Lubimova (1969) concludes that in the continental areas of the U.S.S.R. many of the heat flow variations are due to the tectonics of a given area. Lee (1965), in a summary of more than 3,000 heat flow measurements, concludes that heat flow is generally low in Precambrian shields, average in surrounding cratonic areas, and high in many Mesozoic-Cenozoic orogenic areas at the continental margins.

Sass (1971) concludes that observed heat flows are generally compatible with the hypothesis of sea floor spreading and plate tectonics, that is, high along ridges where hot material is supposed to be rising and low near trenches where a cold plate evidently is descending (Fig. 8).

TEMPERATURE AND HYDROCARBON FORMATION
AND ACCUMULATION

Heat flow is determined from the temperature gradient observed in boreholes or probes, in comparison to the thermal conductivity of the rocks involved. This is only an approach to the true heat flow, which also involves losses by convected waters (see Morgan, this volume) but is the term normally reported. A rough way of measuring the thermal gradient is to observe bottom hole temperatures and subtract average surface temperatures and calculate the gradient in degrees Fahrenheit per 100 feet of depth (degrees Celsius per 100 meters).

Landes (1967) estimates that worldwide geothermal gradients range between 1.0 and 1.6°F/100 ft (1.8± and 2.9±°C/100 m) using deep wells below 15,000 ft (Fig. 2). As an average, 1.4°F/100 ft (2.5±°C/100 m) could be used, although extremes of 42°F/100 ft (65.4±°C/100 m) in the thermally active Imperial Valley (Meidav, 1970) to 0.3°F/100 ft (0.6±°C/100 m) in the Bahamas (Landes, 1967) have been recorded.

Temperatures at depth affect the source, reservoir, and cap rocks involved in the formation, migration, and entrapment of hydrocarbons. Hunt (1962), Welte (1965), and Tissot (1971) have noted that the release of fatty acids or lipids from kerogen (that is, the hydrocarbon base material from decayed organic matter) is a temperature-related process. This cracking process is carried out at depth, and both temperature and pressure effect the rate of oil formation. Of the two, Phillipi (1965) indicates that temperature is the most important, generally increasing the chemical reactions of hydrocarbon formation exponentially (Fig. 1). Levorsen (1967) has described the heat effects on hydrocarbon viscosity, volume, pressure, and solubility. Higher temperatures lower the viscosity and increase the volume, pressure, and solubility.

Most authors generally agree, and considerable field evidence indicates that shales are among the sedimentary materials richest in organic matter—the source rock for hydrocarbons. Following pioneering work by Hedberg (1926, 1936) and Athy (1930) on the compression of shales (simulating depth of burial), present analysis indicates at least three phases of dehydration. Burst (1969), using Gulf Coast data, has constructed a model of shale compaction (Fig. 3). The first phase involves "squeezing" or discharge of pore water by compaction resulting from overburden pressure. A second dewatering phase appears to be related to thermal activity and possible clay transformation—as in the dehydration of montmorillonite (Karstev, 1971)—and results in the molecular discharge of water from clay interlayers. Chilingar (1960) has shown that montmorillonite contains and retains more water than illite and kaolinite under compaction in laboratory experiments. The third phase is slow dehydration with little water or liquid hydrocarbon movement. During this phase, the density of the shales and their relative impermeability is increased.

Recently Perry and Hower (1972) have introduced a threefold subdivision of Burst's stage I and II and emphasized the role of pressure and pore water composition. However, all authors tend to agree that in a high geothermal gradient dehydration occurs at relatively shallower depths than in a low geothermal gradient.

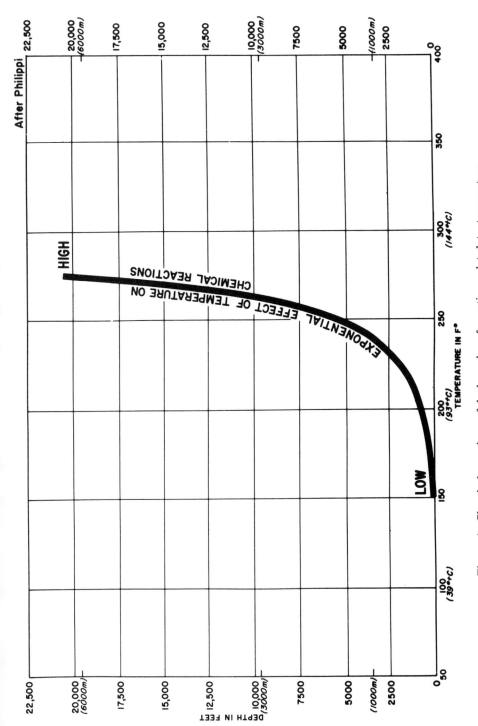

Figure 1. Chemical reactions of hydrocarbon formation related to temperature.

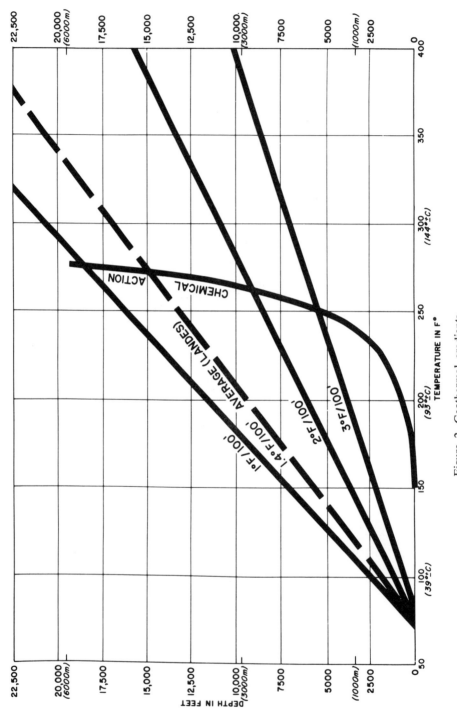

Figure 2. Geothermal gradients.

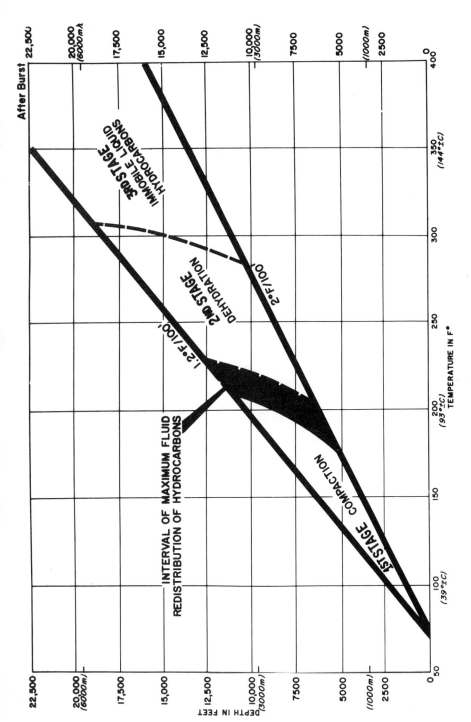

Figure 3. Shale fluid distribution model (Gulf Coast).

Based upon findings in the Gulf Coast, Burst (1969) has defined an interval in which maximum redistribution of pore water and hydrocarbons is most likely to occur. He documents this with field evidence by plotting the depth of production of the majority of the producing horizons in the Gulf Coast and assumes that hydrocarbon movement was initiated mainly at the end of stage I and the beginning of stage II of his three stage dehydration model (Fig. 4). In this phase, the hydrocarbons were then transported upward or laterally to the nearest reservoir and trap. It has been assumed that the petroleum moved either as globules or in micelle form (that is, as colloidal particles or like a soap suspension) during primary migration (Welte, 1965). Karstev (1971) and other Russian investigators concur with these findings and conclude that the "principal stage in the formation of petroleum corresponds with the reconstruction of clay minerals when large volumes of interstitial waters are released." They have termed this phase *katagenesis*. It corresponds to the second stage of dehydration and interval of maximum fluid redistribution as outlined by Burst. They indicate that it occurs in a temperature range of 140°F to 266°F (60°C to 130°C).

It is estimated that upwards to ⅔ of the world's oil and gas reserves are in sandstone, with the major portion of the remaining reserves in carbonates (Halbouty, et al., 1970, and Klemme, 1971). Maxwell (1964) has found a linear, rather than exponential, decrease in porosity of sandstone with depth (Fig. 5). He relates this to pressure from overburden compaction and increased solubilities due to temperature. The geothermal gradient has some effect on reservoirs, namely, a slight increase in the linear porosity gradient (see Tertiary example). However, the decrease is essentially linear, at least above 25,000 ft. Time was also found to have a definite influence on porosity (see Tertiary vs. Paleozoic, Fig. 5). Walker (1964) suggests that this decrease in porosity may be accelerated below 25,000 ft. The individual and the averaged ages of Maxwell's samples are plotted on the diagram. The average selected is an optimum, and production in the field often comes from reservoirs with a much lower porosity spread.

To summarize: the effect of temperature on the chemical reactions involved in petroleum genesis is exponential, and high temperatures aid the expulsion of fluids from shale source beds. On the other hand, the loss of porosity in reservoirs—at least in those of sandstone type—is linear with depth. This implies not only that relatively high temperatures favor the generation and migration of petroleum but also that a high

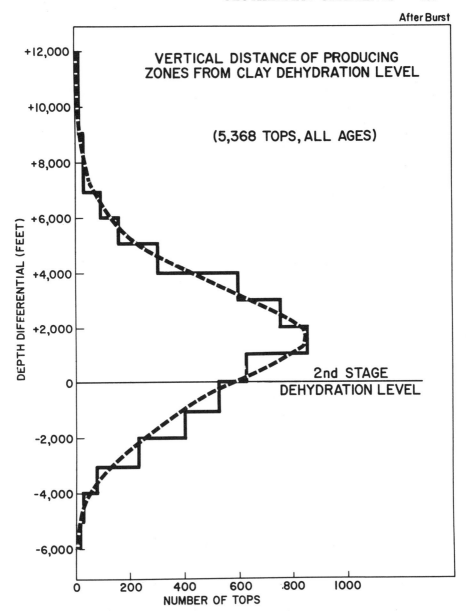

Vertical distance of productive zones from clay-dehydration level, Gulf Coast area, using data from 5,368 production levels.

Figure 4. Gulf Coast productive zones related to shale fluid distribution model.

thermal gradient facilitates these processes at shallow depths at which reservoir availability is optimal. (Fig. 6).

To examine some specific examples: at 250°F (121°C) the difference in reservoir pore space between a normal gradient and a high gradient of 3°F/100 ft (5.5 ± °C/100 m) is nearly 50%; i.e., 38% porosity in the average sandstone at 250°F (121°C) with a 3°F/100 ft (5.5±°C/100 m) gradient versus 26% porosity at 250°F (121°C) with a 1.4°F/100 ft (2.5±°C/100 m) gradient. If the maximum fluid redistribution zone is used with a geothermal gradient of 3°F/100 ft (5.5±°C/100 m), oil will enter rocks at 3700 ft with porosities of 42% or over 60% more pore space than in the maximum redistribution zone with a 1°F/100 ft (1.8±°C/100 m) geothermal gradient at depths of 13,000 ft (Fig. 6). Given similar traps, a high geothermal gradient might more than double the petroleum accumulation developed in a low one, due not only to reservoir pore space but to decrease in viscosity, increase in fluid pressures, and higher permeabilities in carrier beds. In addition, once the petroleum is in a reservoir, the normal decrease in pore space will be inhibited by the presence of the petroleum. Thus, given all the necessary parameters for oil formation and entrapment, the basin with a relatively high temperature gradient will be the one most likely to yield major petroleum accumulations.

In many respects, recent findings by Brooks (1970 and 1971) relating coal deposits to geothermal gradients reconfirm Hilt's law (Emmons, 1931) that not only does the degree of carbonization of coals normally increase with depth but that it is also temperature related. These data also, with few exceptions, generally fit the index of White's (Emmons, 1931, and Levorsen, 1967) carbon ratio theory. Landes (1967) and Brooks (1970) separately have constructed models classifying various coals and oil and gas combinations utilizing the effect of temperature (Fig. 7). Using the geothermal gradients of 12 Mesozoic or Tertiary giant fields Pusey (1973) has proposed a "Liquid Window Concept" wherein with increasing temperature (thermal gradient) a progression from biogenic gas to oil to thermal gas occurs. The liquid or oil phase occurs between 150°F to 300°F at depths ranging from 16,000 to 2,500 feet depending on the geothermal gradient. It is possible that desulfurization and increases in hydrogen sulfides may show some relation to temperature (Vredenburgh and Cheney, 1971). Many low-sulfur crudes in some basins come from deeply buried units whereas the same units at shallow depth produce high sulfur crudes. Thus the depth at which

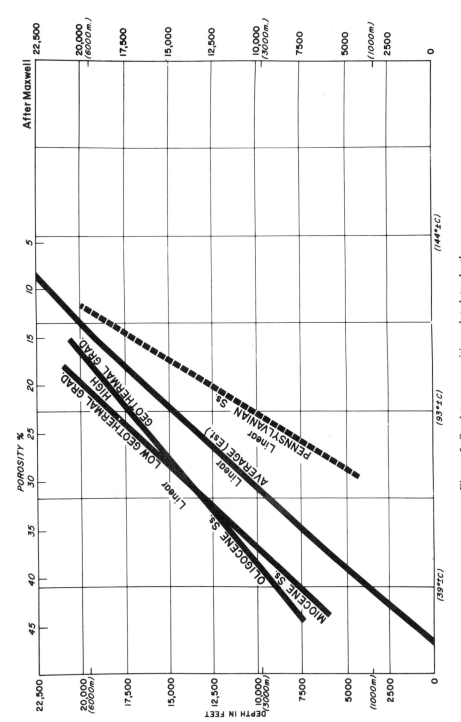

Figure 5. Sandstone porosities related to depth.

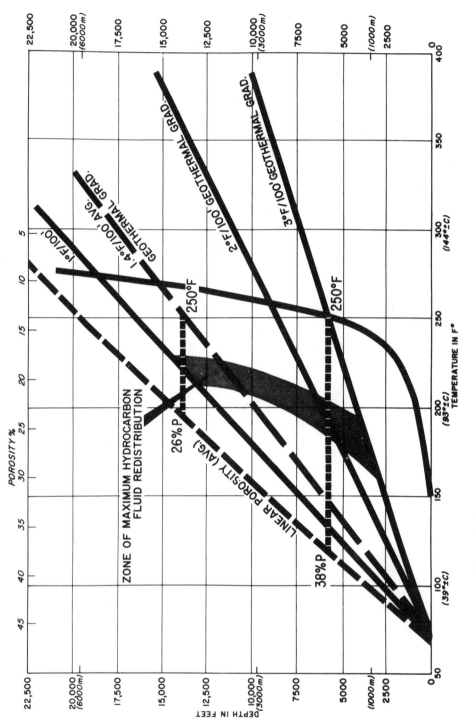

Figure 6. Model relating chemical reaction in clastic sediments and hydrocarbons to geothermal gradients and porosity changes in sandstone.

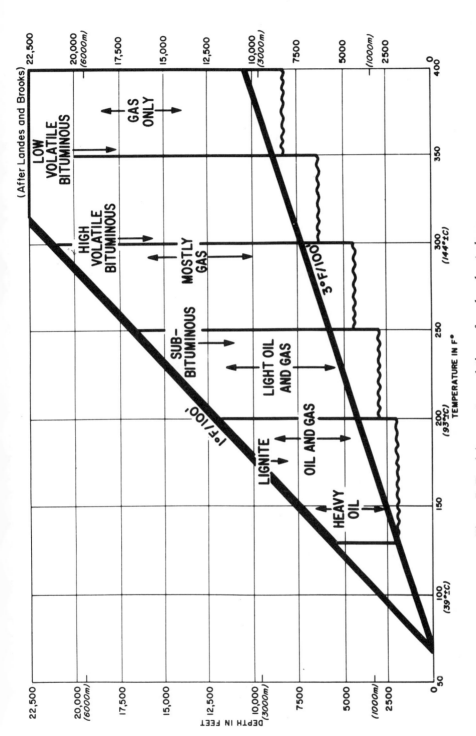

Figure 7. Depth/temperature relations of coal and petroleum.

these temperature-dependent changes occur will be greatly affected by the geothermal gradient.

Heat Flow, Geothermal Gradients, and Hydrocarbon Accumulations in Various Basin Types

Turning to field examples, we may try to relate these physical parameters to petroleum basins in light of plate tectonics. The basins that contain giant fields of over 500 billion bbls. of oil or gas equivalent (Klemme, 1971) provide a sample that includes more than 75% of the world's reserves. One of several possible tectonic classifications has been outlined. Many of the world's hydrocarbon-bearing basins may be related to sea floor spreading (Fig. 8). The classifications used herein are based, in part, upon the concepts of plate tectonics. Within the continents, major tectonic zones include the Precambrian shield areas surrounded by sediment-covered sialic crustal zones of the craton. Along the continental margin, an intermediate crustal zone (Hedberg, 1970) lies between the oceanic crust with an average thickness of 3 miles and the cratonic crust with an average thickness of 20 miles. Variations and gradations in crustal thickness occur most commonly between these two types of crust. This intermediate zone commonly shows a significant divergence between the trends of the younger Mesozoic and Tertiary orogenic zones and older Paleozoic ones (Fig. 9). Paleozoic fold belts appear, in general, to occupy concentric positions around Precambrian shield areas, whereas Mesozoic and Tertiary fold belts parallel subducting or underthrusting oceanic margins.

One possible step in relating the geothermal gradient to petroleum formation is to examine the depth at which oil presently occurs within the various types of basins and to relate this depth to temperature. This is done in a fashion similar to Burst's (1959) compilations in the Gulf Coast although a greater grouping in the range of depths has been plotted. This plot is based upon the reserves of the world's giant fields.

Of major hydrocarbon deposits around the world, 48% lie above 6,000 ft, 43% between 6,000 and 9,000 ft, and 9% below 9,000 ft (Fig. 10). To some extent the high percentage of oil reserves in the zone above 6,000 ft represents the early history of petroleum development before and immediately after World War I, when drilling was limited to shallow depths. Between 1956 and 1970, more reserves were

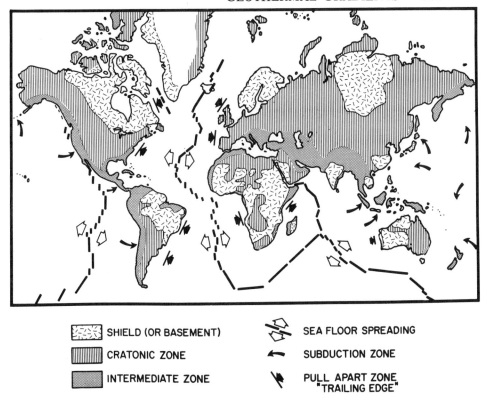

SHIELD (OR BASEMENT)   SEA FLOOR SPREADING

CRATONIC ZONE   SUBDUCTION ZONE

INTERMEDIATE ZONE   PULL APART ZONE
"TRAILING EDGE"

Figure 8. World basin zones related to sea floor spreading.

added to the 6,000 to 10,000 ft zones of production than were added to zones above 6,000 ft. This fact has lead many to conclude that major petroleum deposits will be found by simply drilling deeper in many basins. The assumption may be incorrect in many types of basins. In fact, in many basins with a development history of over 50 years (including deep drilling) 75% of the reserves in giant fields were discovered during the first 10 years of development (Klemme, 1971b). In these instances, deep drilling has not added any substantial reserves. The temperature-related formation of oil and gas as outlined by Landes and Brooks suggests an increasing percentage of gas to oil will be present at depths below 12,000 ft. It therefore seems doubtful that the amount of oil to be found below 9,000 ft will be equal to that found between 6,000 and 9,000 ft.

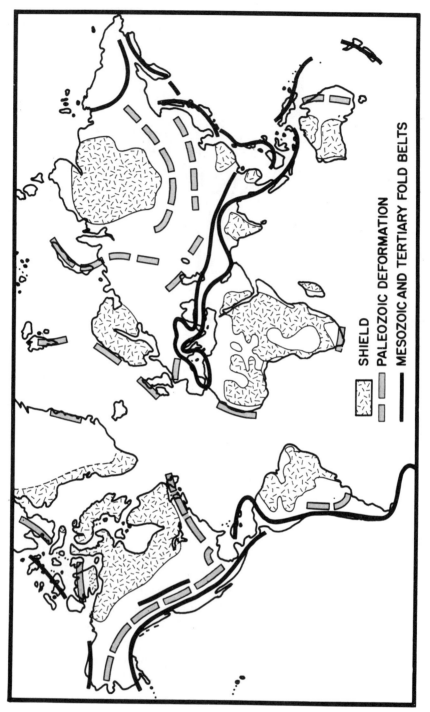

Figure 9. Paleozoic versus Mesozoic and Tertiary deformation belts.

SHIELD

PALEOZOIC DEFORMATION

MESOZOIC AND TERTIARY FOLD BELTS

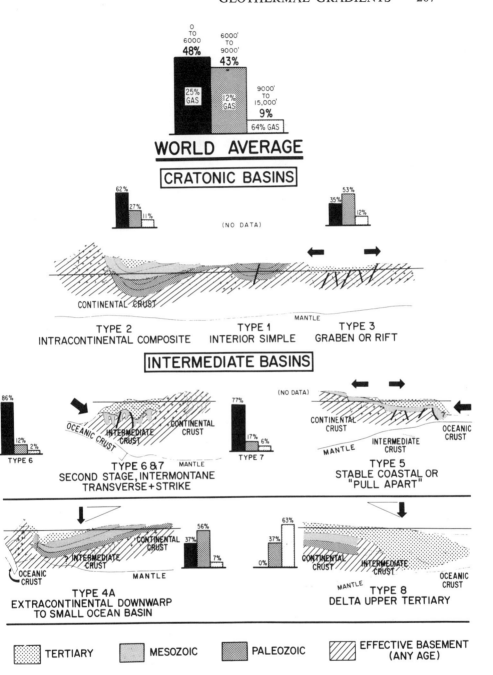

Figure 10. Depth of principal hydrocarbon reserves related to basin type.

Petroleum Basin Classification (Klemme 1971)

Petroleum basins may be classified on the basis of their regional tectonic character and by their relation to continental and oceanic crust. The classification outlined here has evolved from the tectonic types defined by Weeks (1952) and Uspenskaya (1967) and then related to the concept of sea floor spreading. For purposes of this discussion, eight major basin types are recognized. Three of these types are developed on continental crust. Five are developed in the "intermediate" crustal zone between the oceanic crust which averages 3 miles in thickness and the continental crust which averages 20 miles in thickness. In addition to the eight basin types outlined, basins underlain by oceanic crust exist but are presently beyond the reach of present economic operations.

Although cratonic basins (i.e., basins underlain by continental crust) predominate in continental areas, some cratonic basins may extend offshore for considerable distances (North Sea, Pechora, Aquitaine), whereas intermediate basins (i.e., basins underlain by crust intermediate to that beneath continents and that beneath oceans), though generally associated with continental margins, may be found in areas now considered continental (San Joaquin, South Caspian-Baku areas, Maracaibo, Vienna basin) (Halbouty, et al., 1970). The juxtaposition of continental and intermediate crustal types of basins is compatible with the concept of sea floor spreading.

Cratonic Basins

TYPE 1. INTERIOR (SIMPLE)—MODERATELY LARGE, FLAT, SAUCER-SHAPED, SINGLE CYCLE BASINS

These basins are located in the interior of continental areas near Precambrian shield areas. Their sediments often consist of Paleozoic platform or embayment facies. Basement uplifts and sedimentary structures make up the petroleum traps. These basins generally yield high gravity, low sulfur crudes.

TYPE 2. INTRACONTINENTAL (COMPOSITE)—LARGE (SUBCONTINENTAL MIOGEOSYNCLINAL) TO SMALL (INTERMONTANE) BASINS WITH TWO OR MORE CYCLES OF DEPOSITION

These basins are less simple, multicycle basins (usually orogenic Mesozoic sediments over Paleozoic platform sediments) located in the more exterior parts of continental cratonic areas. They generally yield high gravity crude with low sulfur in clastic reservoirs and high sulfur in carbonate reservoirs. Over 75% of the world's gas is found in these basins. Evaporites are often present and field sizes appear to depend on whether a large regional arch acts as a trap; in which case up to 75% of the basin's reserves are in one field (if not, field sizes are less than 10% of the basin's reserves in small stratigraphic and structural-stratigraphic traps).

TYPE 3.   RIFT (GRABEN OR HALF GRABEN)—SMALL TO MEDIUM
SIZED, DOWNFAULTED BASINS

These basins, located on continental crustal zones, often extend toward the continental margin. They appear to result from incipient sea floor spreading (rift on stable craton) in post-Paleozoic time. In some instances they are developed upon earlier Type 2 basins (i.e., North Sea Tertiary graben). The sedimentary facies control hydrocarbon types which are generally high gravity, low sulfur crudes. Evaporites appear to enhance hydrocarbon recovery.

Intermediate Basins

TYPE 4.   EXTRACONTINENTAL (DOWNWARPS INTO SMALL
OCEAN AREAS)

These basins are located along the continental margin and extend into the adjacent small oceans (Gulf of Mexico, Tethyan, Caribbean, China and Beaufort Seas). They are predominantly post-Paleozoic in age, and some have been referred to as "successor" basins (King, 1969) and epieugeosynclines (Kay, 1951). Several subtypes are present in this category. Intermediate gravity and higher sulfur crudes are present in these basins.

TYPE 5.   PULL-APART—LINEAR, ONE-SIDED, DOWNFAULTED
BASINS OF STABLE COSTS

These basins are located along continental margins (particularly of the Atlantic and Indian Oceans) and consist of downfaulted Mesozoic and Tertiary sediments. They appear to be Type 3 (rift) basins that have been separated to oceanic distances by sea floor spreading ("trailing edge" of continental blocks). To date they have had little exploration activity offshore and both their potential for petroleum and their hydrocarbon character are poorly known.

TYPE 6 AND 7.   INTERMONTANE SECOND STAGE BASINS

These basins are located along continental margins (where an oceanic plate appears to underthrust a continental plate). They are Cretaceous and Tertiary in age, consisting of terrigenous clastic sediments resulting from a second cycle of deposition over deformed and intruded Mesozoic eugeosynclines previously developed parallel to the continental margin. If favorable geologic conditions (facies, structure, etc.) are present, prolific production often is present. Hydrocarbon types are variable although intermediate gravity, high sulfur crudes low in gas predominate. Type 6 basins lie transverse to margin, type 7 basins parallel.

TYPE 8.   UPPER TERTIARY DELTA—UPPER MIOCENE TO RECENT, BIRDFOOT DELTAS ALONG CONTINENTAL MARGIN WHERE MAJOR CONTINENTAL DRAINAGE AREAS EXIST

These basins contain young topset, bottomset and forset, marine and nonmarine clastics with "rollover" and salt flowage structural traps. Generally low sulfur, high gravity crudes are present, and gas occurs in abnormal amounts. The field sizes are generally less than 5% of the delta's total reserves.

The effect of the geothermal gradient on petroleum occurrence appears to be greatest in the intermediate crustal zone basins that occupy the transition between continental and oceanic crust. These basins are located most often along the margins of major plates and account for over two-thirds of the world's hydrocarbons. Because the intermediate zone contains the continental shelf, it is expected that more than two-thirds of the future reserves will come from these basins (Weeks, 1965, 1971, and Klemme, 1971b). Whereas among the cratonic basins only Type 3, or rift basins, are direct expressions of plate motion, all of the intermediate basins have a more or less direct relation to this geologic process. Nearly all the reserves of the intermediate basins are of Mesozoic or Tertiary age and are predominantly in clastic rocks.

To illustrate the action of the geothermal gradient, we may start with the late Tertiary deltas. These deltas are generally tectonically simple being unidirectionally downwarped into the ocean areas. They most often have normal geothermal gradients, near the surface. Drilling to depth has shown both abnormally high pressures and abnormally high temperatures, suggesting that the oversaturated shales at depth are preventing the heat from freely flowing to the surface (Lewis and Rose, 1969, and Jones, 1969). These shales appear to act as a "damper," creating a time lag in the ultimate surface heat flow.

In such deltas a great percentage of the major reserves lie at depths below 9,000 ft (Figs. 11 and 12). Immediately offshore in the Mississippi Delta a "hot belt" is present at depths below 9,000 ft, and this zone contains the giant fields of the delta (Jam, Dickey, and Tryggvason, 1969). This east-west "hot belt" may be associated with the introduction of a massive diapiric salt zone. Heat conductivity is more efficient in anhydrites and salt (Zierfuss, 1969), which may enhance oil generation and accumulation.

Basins of Types 6 and 7 (Fig. 10) are associated with belts along which crustal plates are underthrusting or overriding one another (Caribbean area and eastern Asia, Figs. 13 and 14) along subduction zones. The basins lie on overriding (continental) plates behind such subduction zones, and represent second-stage clastic deposits of Cretaceous or Tertiary age, lying uncomformably on older, deformed or even metamorphosed eugeosynclinal troughs. These older troughs parallel the plate margin (Halbouty, 1970), whereas the young basins developed over them may lie transverse (Type 6) or parallel (Type 7). The heat flow in these basins is generally high, and the petroleum reserves are located at shallow depths.

In southeast Asia, three Type 7 basins lie along the strike of the classic Indonesian subduction zone, more specifically, behind the Indonesian volcanic arc (Fig. 15): the Central Sumatra basin, the South Sumatra basin, and the Sunda basin—the Java Sea. All have rather similar sediments, structural traps and tectonic origin. While details of heat flow have been released only for the Sunda basin, very high to extremely high gradients are known to exist in Central Sumatra, particularly in the vicinity of the supergiant Minas field, where gradients of $4°F/100$ ft ($7.3±°C/100$ m) are not uncommon. Production comes from 2500 ft±. More average or normal gradients are reported in South Sumatra. The recovery of oil may be related to the geothermal gradient, with large reserves per cubic mile of sediment in the relatively "hot" Central Sumatra basin and lesser recovery rates in south Sumatra. The Java Sea has not been developed to the point of determining the ultimate reserves per cubic mile; however, high gradients appear to be present.

Transverse and strike Type 6 and 7 Intermontane basins are located not only in island arc subduction areas but also along the intermediate

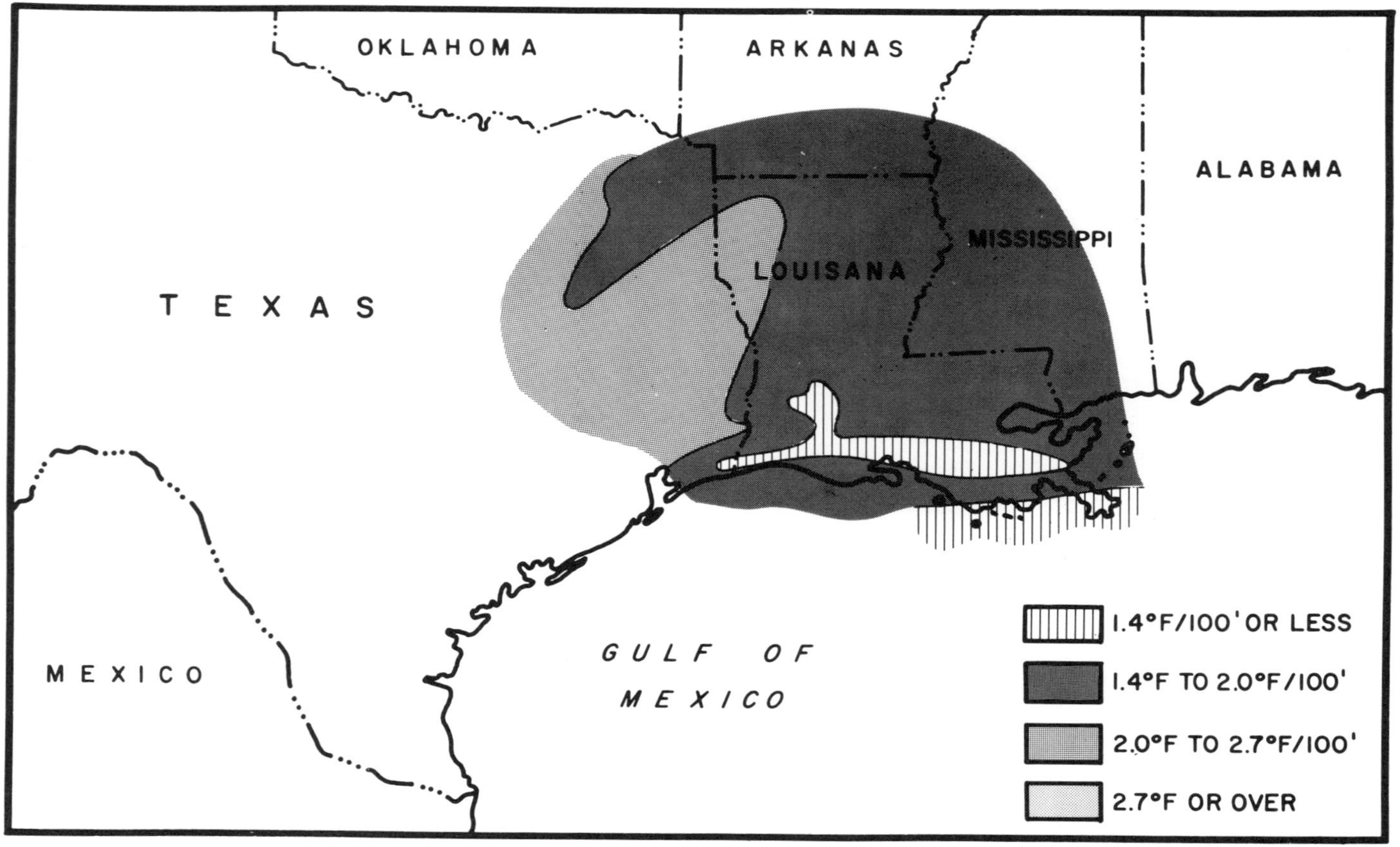

Figure 11. Geothermal gradients in Mississippi delta region.

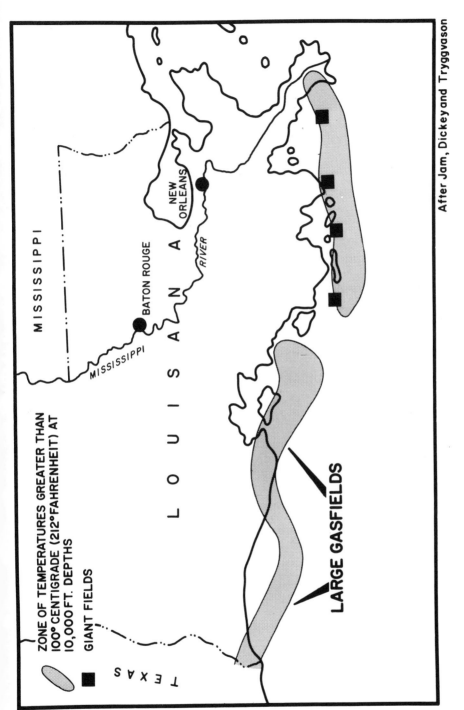

Figure 12. Mississippi delta "Hot Belt" in relation to giant oil fields and large gas producing areas. (Temperatures between 100° and 113°C.)

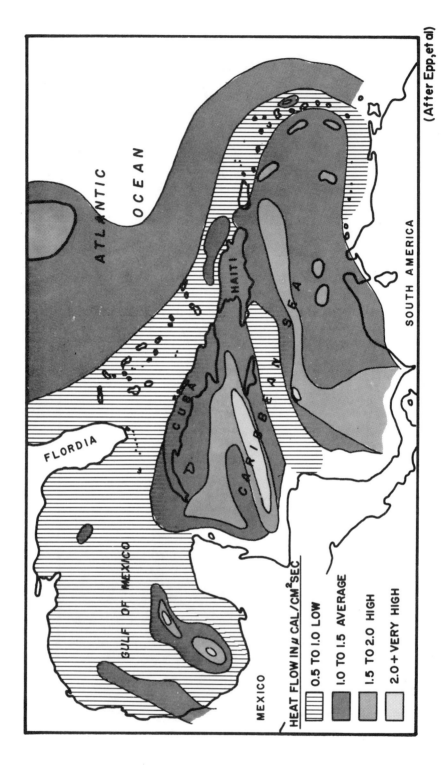

Figure 13. Gulf of Mexico/Caribbean—generalized heat flow.

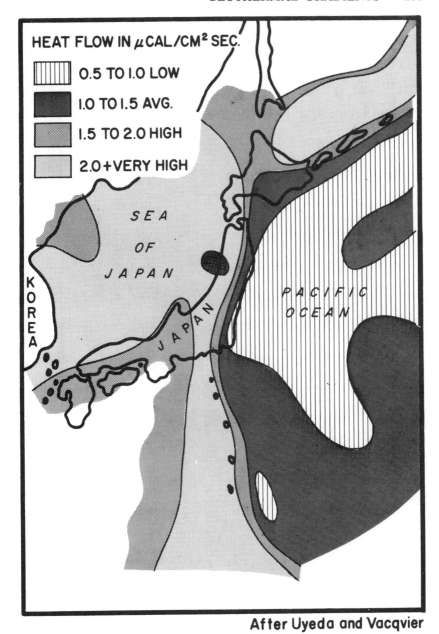

Figure 14. Western Pacific/East Central Asia—generalized heat flow.

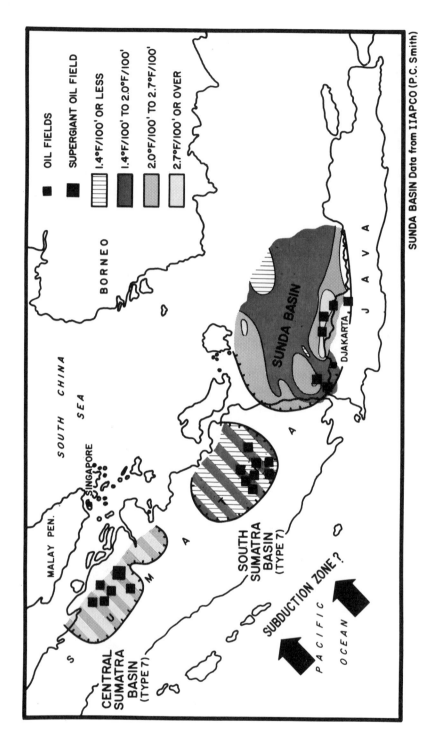

Figure 15. Indonesian island arc area. Zone of subduction with Type 7 intermontane basin development.

zone where continents appear to be overthrusting or overriding oceanic zones as in the western United States. Heat flow (Fig. 16) in this area (Sass et al., 1971) is high along the coast, low in granitic pluton areas such as the Sierra Nevada and Baja California and high in the Gulf of California and possibly central Nevada, where some authors have suggested that spreading may be occurring (Hamilton, 1969, Atwater, 1970; Scholz, 1971).

The Los Angeles and Ventura basins have generally a similar geologic setting but differing geothermal gradients (Fig. 17). In the Los Angeles basin, gradients are high and increase westward to the sea, whereas the Ventura basin has a more normal to low gradient (French, 1939, and Mayuga, 1970). As Phillipi (1965) has pointed out, the high gradient within the Los Angeles basin may account for the more prolific nature of the reserves in this basin as opposed to the Ventura basin. There is not always a relation between present-day high heat flow and correspondingly high geothermal gradients. For example, it may be that the blanket or dampening effect of a thick sedimentary section in the Ventura basin has reduced the geothermal gradients even though the heat input at its base may be as high as the Los Angeles Basin's (Fig. 16 and Fig. 18). In this case, apparently, more rapid deposition occurred in the Ventura basin. A comparison of the geothermal gradient of the two basins is noted in cross sections A-A' and B-B' (data from Mayuga, 1970, and Nagle and Parker, 1971). The giant Wilmington field in the Los Angeles basin has a geothermal gradient of more than $3°F/100$ ft ($5.5 \pm °C/100$ m). The plotted line marks the depth at which $200°F$ ($93°C$) is reached in the two basins. This depth generally correlates with these basins major petroleum deposits (note the oil horizons in Fig. 18). However, oil in substantial amounts does occur at both lower and higher temperatures at lesser and greater depths. Although both basins contain large amounts of oil, yield estimates for the Los Angeles basin are four million barrels of oil per cubic mile of sediments and for the Ventura basin less than 270,000 barrels per cubic mile (Kilkenny, 1971). A comparison of the major fields in the two basins is included in Figure 19.

High geothermal gradients are noted in many other Type 6 transverse basins. For example, the Baku and Vienna basins are developed where two cratonic plates have become welded by moving over or closing upon an oceanic plate. The Baku basin, has high heat flow, increasing down the plunge into the Caspian Sea (Fig. 20). Heat flow is high in the

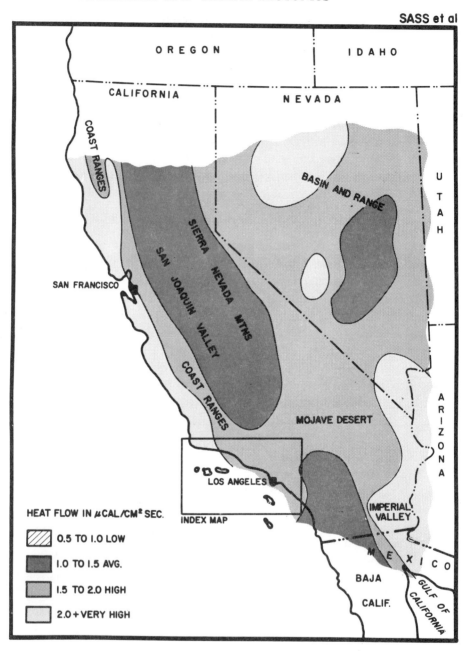

SASS et al

Figure 16. Heat flow in southwestern United States.

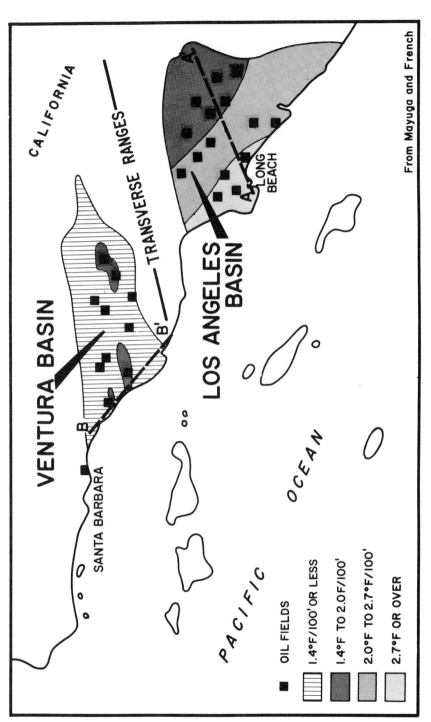

Figure 17. Los Angeles/Ventura basins—geothermal gradients.

LOS ANGELES BASIN

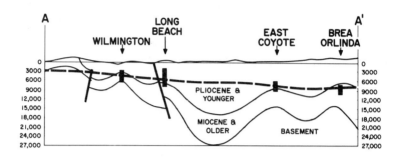

VENTURA BASIN

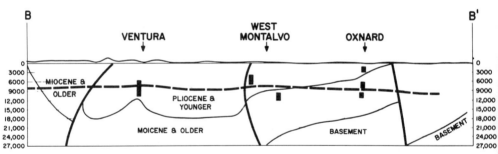

Figure 18. Cross sections of Los Angeles basin and Ventura basin. Dashed line represents 200°F. Production intervals in each field area shown in heavy vertical line. Note higher temperatures in Los Angeles basin and thicker section in Ventura basin.

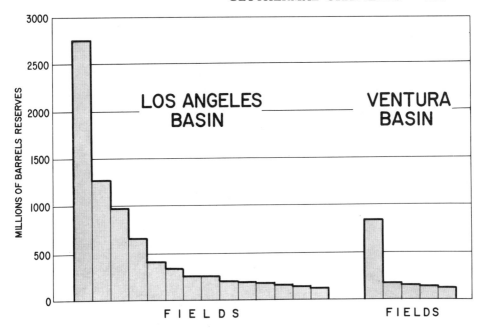

Figure 19. Major oil fields in Los Angeles and Ventura basins.

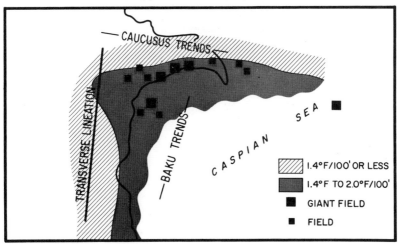

Figure 20. Baku basin, USSR—geothermal gradients.

Vienna basin (Boldizar, 1968), and increases southeast toward the Pannonian basin (Fig. 21). The tectonically similar western Mediterranean basin shows high heat flows (Fig. 22). Both high heat flow and high geothermal gradients are present in southeastern Australia (Fig. 23). The extremely high heat flow and resulting high gradients in the transverse Type 6 Gippsland basin, together with a thicker source and reservoir rock section offshore and downdip, have been an important factor

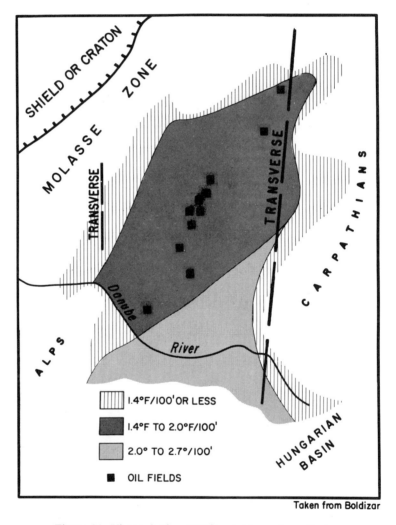

Taken from Boldizar

Figure 21. Vienna basin, Austria—geothermal gradients.

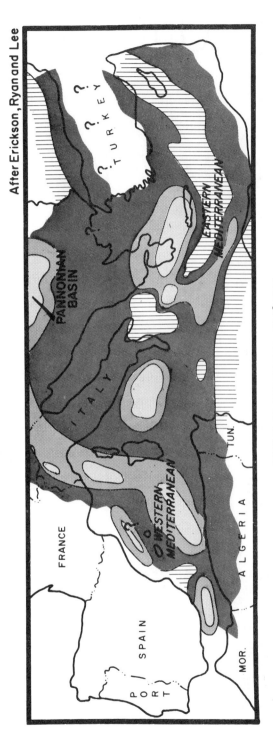

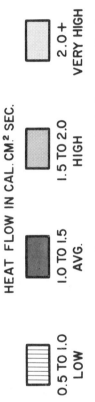

Figure 22. Heat flow of Mediterranean and Pannonian basins (generalized).

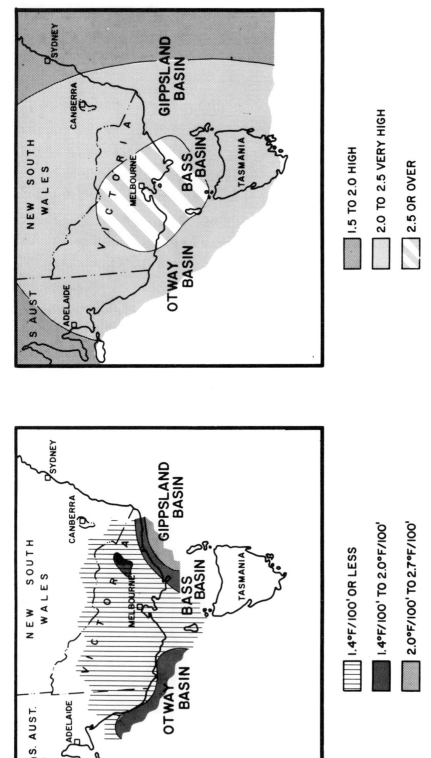

Figure 23. Southeast Australia—Otway, Bass and Gippsland basins—geothermal gradients (on left) and heat flow (on right).

in the formation of several giant oil and gas accumulations (Brooks, 1970b).

Local geothermal "hot spots" have been reported in the San Joaquin, Sacramento, and in other Type 7 basins.

These intermontane Type 6 and 7 basins, generally of Tertiary age, have been tectonically active. Their high geothermal gradients appear to be related to the high heat flow observed behind subduction zones in general. Although relatively high geothermal gradients produce giant accumulations at shallow depths, *extremely* high gradients may destroy the hydrocarbons or at the best reduce the hydrocarbons to gas: (it is suspected that this may be the case in the Pannonian basin in which gas predominates.)

Type 5 or pull-apart basins appear to be the "opening" or an extension of Type 3 rift basins that have been separated by oceanic dimensions. They are present in the trailing edge zones (Fig. 8) along opposite sides of the Atlantic, the Greenland/Canadian Arctic, along the coasts of East Africa, India, and portions of western, southern, and possibly eastern Australia. Because little exploration activity has occurred in these areas, other than in their onshore margins, we have insufficient data to draw any conclusions. Considerable data will no doubt be furnished by the present acceleration of activity in these basins. Theoretically they may be expected to have had high geothermal gradients at the inception of sea floor spreading and lower gradients after the sea floor was spread to oceanic distances.

Type 4 basins are extracontinental downwarps into small ocean basins. They are large to medium in size and located along the margins of present or past small ocean basins. Three types are distinguished: platforms and foredeeps (shown in Fig. 10), foredeeps only, or simple platform downwarps (Klemme, 1971). Producing horizons range from lower Mesozoic to Miocene. Reservoirs are dominantly sandstone except for a nearly 50/50 sandstone/carbonate ratio in the Persian Gulf. The Gulf Coast and East Texas basins represent simple platform downwarps into the small ocean basin of the Gulf of Mexico. Rapid Tertiary submergence of the Gulf Coast and the temporary insulating effect of undercompacted shale has resulted in high geothermal gradients located inland and lower gradients near the coast. The undercompacted shales have acted as a temporary depressant to heat flow along the coast. Inland, however, the undercompacted zone is absent and gradients are higher. In this high gradient zone or adjacent to it lies the supergiant wedge-edge, stratigraphic trap of the East Texas field. Updip lies the giant Carthage gas field (Fig. 24). In Alaska's North Slope basin, gradients of 2.5°F/100 ft (4.5±°C/100 m) have been recorded in the vicinity of the supergiant structural-stratigraphic trap of the Prudhoe Bay field. The Precaucasus/Mangyshlak basin represents a platform and foredeep type of basin (Fig. 25). High to very high geothermal gradients are present along this basin trend. It is presumed that this was a down-warped basin on the Eurasian plate as Tethys closed. Both the Stavropol giant gas fields and the Groznyy giant oil field produce from the perimeter of a high gradient area (for locations see Halbouty, 1970). At Stavropol a gradient of over 2.3°F/100 ft (4.2±°C/100 m) is present.

In the East Venezuela basin, the giant Officina field has a gradient of 2.1°F/100 ft (3.8±°C/100 m). In the Mexican Tampico area along the Gulf of Mexico, the shallow production of the Giant Ebano-Panuco field has a very high geothermal gradient associated with late Tertiary igneous intrusions. These igneous bodies not only appear to have formed some of the structural traps by intrusive doming but seem to have resulted in erratically high geothermal gradients that have been effective in generating oil at shallow depths and moving it into highly fractured carbonate reservoirs (Emmons, 1931).

In the Persian Gulf or Arabian/Iranian basin, normal gradients on the edge of the Arabian shield change to above-average gradients in some of the Arabian producing areas (Fig. 26). However, as the central Iranian massif or nuclear area is approached, low gradients with local "hot spots" are present. Clearly some of the prolific production in the Arabian/Iranian basin is associated with relatively high gradients; however, just as much or more production is associated with low gradients. Therefore temperature does not appear to be the principal factor in the world's greatest producing area.

It appears that in all Type 4 basins, with the exception of the Gulf Coast (with its undercompacted shale at depth), the geothermal gradient is high near the small ocean areas and decreases towards both the craton or shield areas and some of the central massifs in Tethys. Type 4 basins on average have the majority of their reserves at intermediate depths. This average is the same whether the prolific Persian Gulf fields are used or removed from the statistical sample. These usually large basins appear to have above-average gradients and local "hot spots," but the extremely high gradients characteristic of some of the more prolific Type 6 and 7 basins are absent. These basins have temperature characteristics which are compatible with their reserve/depth relations.

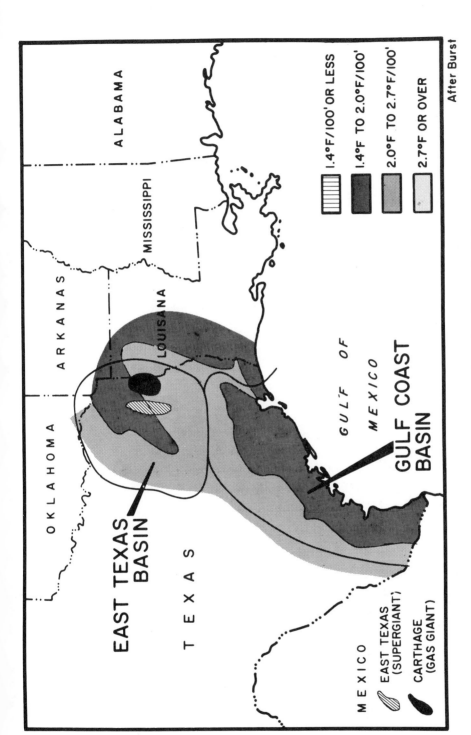

After Burst

Figure 24. Texas Gulf Coast showing geothermal gradients and relation to East Texas and Carthage fields.

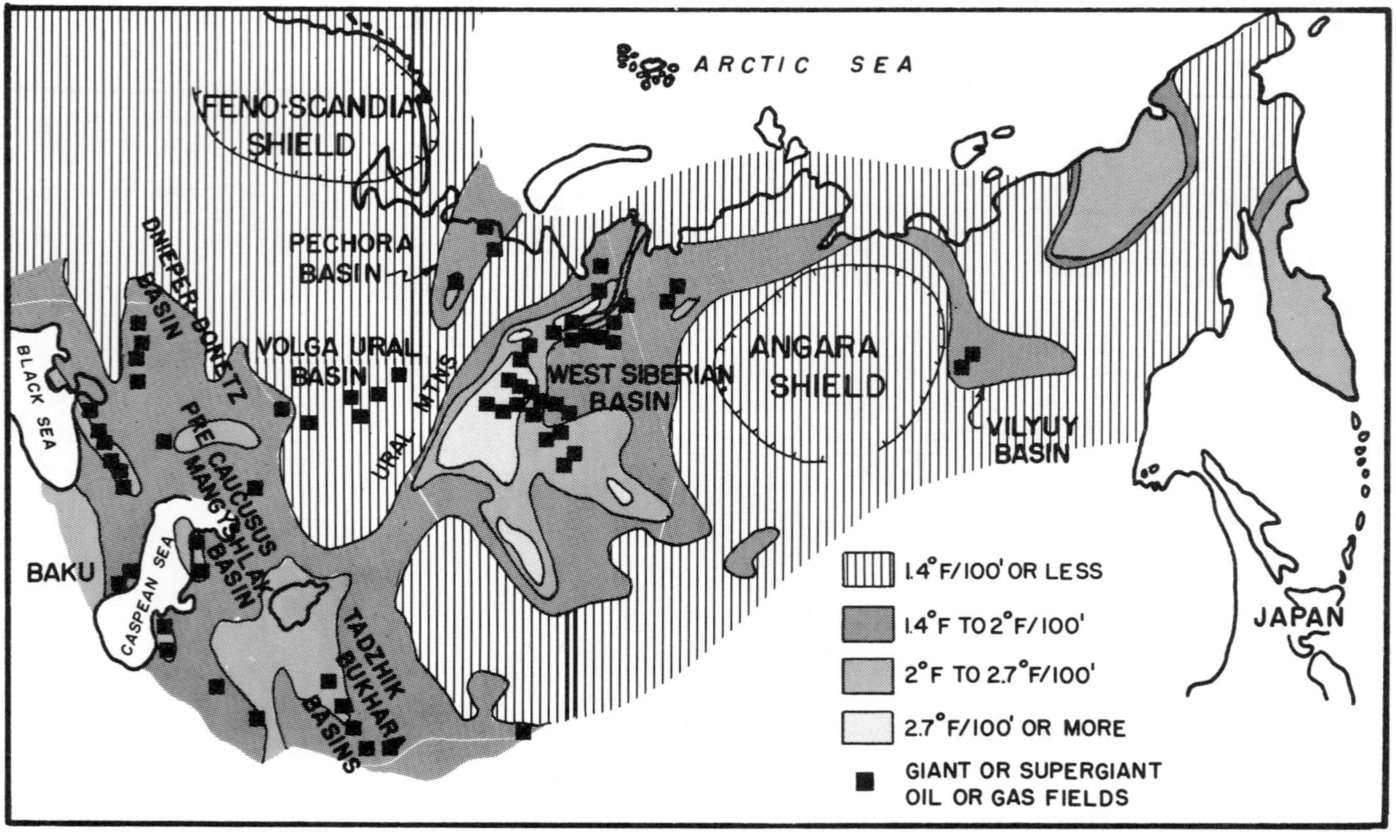
ARCTIC SEA
FENO-SCANDIA SHIELD
PECHORA BASIN
DNIEPER-DONETZ BASIN
VOLGA URAL BASIN
URAL MTNS
WEST SIBERIAN BASIN
ANGARA SHIELD
VILYUY BASIN
BLACK SEA
PRE CAUCUSUS
MANGYSHLAK BASIN
BAKU
CASPEAN SEA
TADZHIK BUKHARA BASINS
JAPAN
1.4°F/100' OR LESS
1.4°F TO 2°F/100'
2°F TO 2.7°F/100'
2.7°F/100' OR MORE
GIANT OR SUPERGIANT OIL OR GAS FIELDS

From Trans. 218 Heat Regime of the U.S.S.R. Interiors-Moscow 1970

Figure 25. USSR—geothermal gradients and hydrocarbon basins.

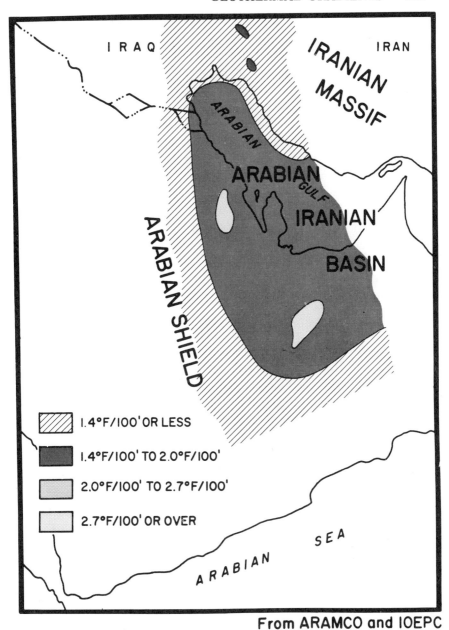

From ARAMCO and IOEPC

Figure 26. The Middle East—Arabian/Iranian basin—geothermal gradient.

The cratonic basins (Types 1 and 2) generally contain high gravity, low sulfur crude. They are estimated to contain over ¾ of the world's gas reserves and 90% of the hydrocarbons that have been found in Paleozoic rocks. With some exceptions, Type 2 cratonic basins are areas of relatively low heat flow and normal to subnormal geothermal gradients. Examples of such basins are the Permian basin of west Texas and southeast New Mexico, the Volga/Ural, Rocky Mountain, Appalachian, Alberta and Anadarko basins. Type 2 intracratonic basins in the Rocky Mountains may be typified by the Powder River and Big Horn basins (Fig. 27). The centers generally have below-average geothermal gradients, while the margins (in this case mountain uplifts) have higher gradients. Other Rocky Mountain basins are similar with subnormal or slightly above-normal gradients in the center of the basin and higher gradients at their outer margins (uncorrected data furnished by members of the AAPG Committee on geothermal gradients). The giant Salt Creek field of the Powder River basin lies in an anomalous zone of higher than normal gradients. It is suggested that much of the hydrocarbons in these mountain basin areas may have accumulated in the vicinity of the Tertiary mountain uplifts only to be stripped and lost by late Tertiary truncation. Similar uplifts with normal geothermal gradients are present in the West Siberian basin. However, in Siberia the arches have not been truncated and therefore yield many giant and supergiant fields. At the same time normal to low heat flow measurements consistent with cratonic areas have been recorded in portions of this basin (Moiseyenko, 1970, and Fotiadi, 1970). Some evidence has been presented that the West Siberian basin, presently classified as a cratonic Type 2 basin, may initially have been a Type 4 downwarp basin that either closed upon (Hamilton, 1970) or opened from (Tamrazyan, 1971) the European shield and platform at the Urals. During the deposition of the sediments that contain the hydrocarbons the basin was essentially cratonic, although it presently has a relatively thin rather than thick crust.

Type 3 or rift basins have the majority of their reserves at intermediate to shallow depths. High heat flow has been recorded in all basins of this type from which data are available. The Red Sea appears to be a Type 3 basin that is in the process of passing to a Type 5 "pull-apart" basin. In the Red Sea, extremely high sea floor temperatures are present with associated hot brines. High heat flow measurements have been recorded in the southern Red Sea (Lowell *et al.*, this volume). Drilling temperatures beneath the thick salt deposits in the south part of the

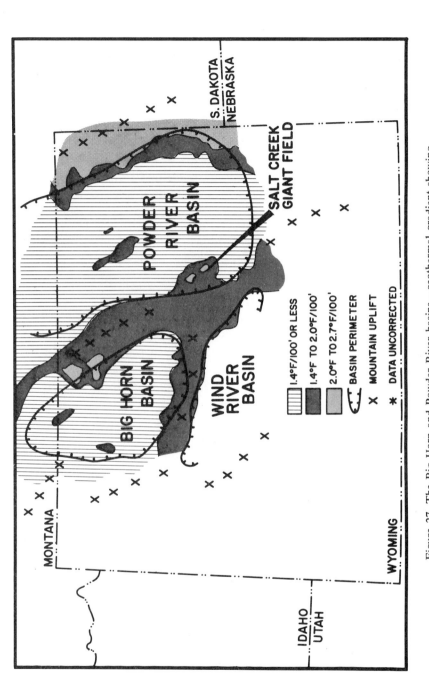

Figure 27. The Big Horn and Powder River basins—geothermal gradient showing relation to mountain uplifts and giant fields.

area have been so hot as to cause serious deterioration of the drill bits. In the case of the Red Sea, the geothermal gradient may be so high locally (such as in the south) as to preclude oil and possibly even gas below 10,000 ft. Similar phenomena and high gradients are reported from the giant African rift. In the Gulf of Suez rift normal to considerably above normal gradients are reportedly present; however, both the heat flow and geothermal gradients do not appear to be as high as in the southern part of the Red Sea.

In the Rhine Graben (Fig. 28) we have an example of a high heat flow with resulting high geothermal gradients; however, the prerequisite of the necessary geologic factors for oil is restricted (Delattre, 1970; Hanel, 1970). Thus, even with high temperatures, large reserves are lacking.

Heat flow in the Indian Cambay area, which is another rift-like basin, is high in several areas. Gupta (1970) explains this as due to igneous intrusions that were emplaced in Miocene and Pliocene times in the crust underneath the basin. In the Dnieper-Donetz and Bukhara Type 3 rift basins (Fig. 25) relatively high geothermal gradients are present (Peive, 1970).

Surprisingly, there are no data to determine whether or not the North African Sirte basin has high heat flow or a high geothermal gradient similar to the other rifts of Africa. One suspects that either presently high geothermal gradients or high paleogeothermal gradients were present, and the extensive evaporite cap rock that covers much of the basin's producing areas may have acted as both a sealing cap rock and a conductor of heat much as the lid on a kettle. Zierfuss (1969) indicates that evaporites are extremely high heat conductors.

In the North American Basin and Range province, which resembles a series of Type 3 rift-like basins, higher than normal heat flow has been recorded (Sass, 1971). This region demonstrates that high heat flow alone is not sufficient to provide hydrocarbons if the geologic parameters essential to the formation, migration, and entrapment of oil were not initially present.

GEOTHERMAL GRADIENT AND DEPTH RELATIONS OF HYDROCARBONS

The depth of major reserves in the various types of basins suggests relation to heat flow and the resulting geothermal gradient. In many

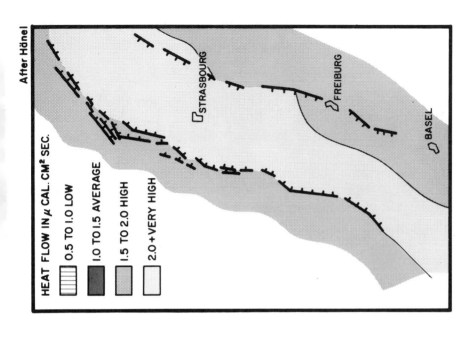

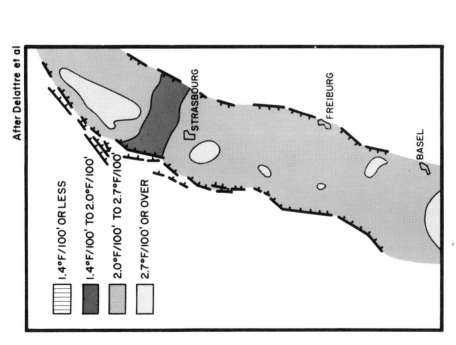

Figure 28. The Rhine Graben basin with geothermal gradient on the left and heat flow on the right.

Type 6 and 7 basins, shallow and prolific oil accumulations appear to have resulted from the early generation, migration, and accumulation of hydrocarbons in shallow reservoirs during the compaction phase, when water was squeezed from shales. Within limits, the higher the gradient, the greater are the chances for giant fields.

In the case of late Tertiary to Recent (Type 8) deltas, the rapid submergence of terrestial and marine sediments has often formed a temporary insulating zone of undercompacted shale that dampens the heat flow at the continental margin. Near-normal geothermal gradients are present, and therefore deeper production with larger-than-normal amounts of gas are present. This is due both to the nature of the source rock material and to the temperature gradient as related to the depth of burial of the sediments.

In the case of Type 5 or pull-apart basins insufficient data preclude any conclusions, although low gradients theoretically might be expected following high paleogeothermal gradients in the initial rifting stage of development.

Type 4 basins have both a high geothermal gradient with zones or areas of "hot spots" near the continental margin that change to normal gradients as either the craton or some of the outer massifs (Tethys) are approached. The high percentage of carbonates in these basins does not always allow the application of the clastic thermal models originally discussed. In spite of the complications of carbonates, it appears that the bulk of the reserves comes from intermediate depths associated with moderately high to normal temperatures along a trough to platform hinge line or where local hot spots occur.

The relation of sandstone and carbonate reservoirs by basin type (Fig. 29) indicates that sandstone-shale production predominates in the intermediate basins, including pull-apart, delta, and subduction zone intermontane types whereas cratonic basins and intermediate Type 4 extracontinental downwarps have a higher ratio of carbonate reservoirs.

In Type 3 basins high heat flow appears to have been present in the initial stages. Continued subsidence and cooling of some of the rifts have left most of the hydrocarbons at intermediate depths.

The high percentage of shallow production in the Type 2 basins appears to be caused by several factors, including the following:

1. The unburdening or unroofing of cratonic basins by truncation. In this case the initial deep burial of the basin with higher paleotemperatures was followed by uplift and truncation thereby lifting the oil deposits, through time, to the shallower depth and lower temperatures

CRUST	CRATONIC			INTERMEDIATE				
TYPE	1	2	3	4	5	6	7	8
CARB	67%	23%	42%	17%	17%	2%	5%	–
Ss	31%	75%	55%	52%	83%	98%	95%	100%

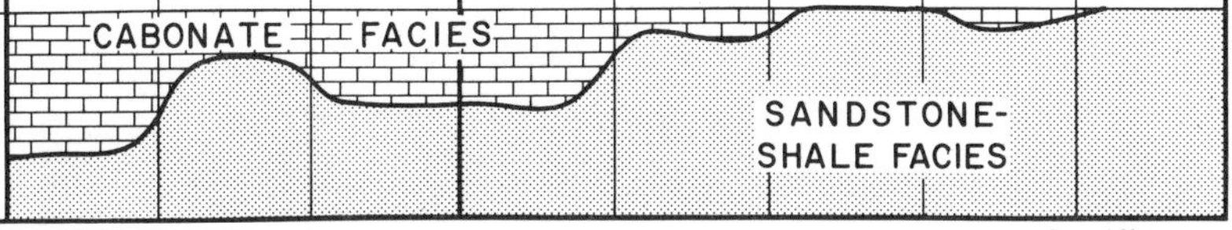

CABONATE FACIES

SANDSTONE-SHALE FACIES

After Klemme

Figure 29. Relation of sandstone and carbonate reservoirs by basin type.

from which they now produce. Evidence for unburdening or unroofing of many cratonic basins is found in the presence of high density shales at shallow depths. High density shales must have been deeply buried to reach their degree of compaction and subsequent unroofing apparently has not completely reversed the density. In addition, the high sulfur content of many Paleozoic crude oils in these basins suggests a limited time of deep burial—else thermal desulfurization would had occurred.

It is possible to measure the organic maturity of any kerogen-bearing rock-type and thereby establish rough paleotemperatures (Pusey, 1973). Present geotemperatures and past pealeotemperatures as determined by maturation analysis are key elements in establishing the source rock potential in new or new parts of basins.

2. Many carbonate reservoirs are present, and the hydrocarbons in them may have had a different history of accumulation than in the clastic sequences considered here.

3. Time or the element of "cooking time" (Dott and Reynolds, 1969, p. 91) has not received proper stress in this study, which is primarily concerned with temperature. However, time appears to have an important influence on petroleum generation in general (i.e., compaction of shale and sandstone, changes in hydrocarbon character (maturation), migration of hydrocarbons, tectonic history including rate of deposition

CRUSTAL TYPES		BASIN TYPE	GEOTHERMAL GRADIENT (NUMBER OF BASINS USED)	RECOVERY BBLS.(OR GAS EQUIVALENT)PER. CU. MI. OF TOTAL BASIN SEDIMENT
CRATON	I.	INTERIOR	4 LOW	18,000
	2.	INTRACONTINENTAL COMPOSITE	I HIGH 12 LOW TO NORMAL (25 UNKNOWN*)	(200,000) 83,000
	3.	RIFT	5 HIGH (3 UNKNOWN*)	120,000
INTERMEDIATE	4.	EXTRACONTINENTAL DOWNWARP	7 NORMAL TO HIGH WITH "HOT SPOTS" (15 UNKNOWN*)	280,000
	5.	PULL-APART	(ALL UNKNOWN*)	?
	6.+7.	INTERMONTANE STRIKE & TRANSVERSE	6 HIGH 3 NORMAL TO HIGH (15 UNKNOWN*)	420,000 (240,000)
	8.	DELTA	2 NORMAL (ABOVE OVER PRESSURED ZONE)	200,000

* UNKNOWN = NO DATA ON GEOTHERMAL GRADIENT   (ESTIMATED)

Figure 30. Comparison of basin types to geothermal gradients and hydrocarbon recovery.

TABLE 1

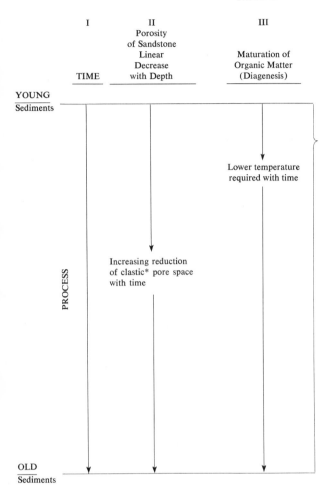

	I	II	III
		Porosity of Sandstone Linear Decrease with Depth	Maturation of Organic Matter (Diagenesis)
	TIME		

YOUNG
Sediments

Lower temperature required with time

PROCESS

Increasing reduction of clastic* pore space with time

OLD
Sediments

MOST FAVORABLE CONDITIONS

YOUNG (TERTIARY OR CRETACEOUS) CLASTIC RESERVOIRS WITH HIGH GEOTHERMAL GRADIENTS

(a) Once oil is in reservoir, it inhibits process in II.

(b) Although more oil is derived in time at lower temperatures (III); less reservoir space in clastics is available (II).

Thus, due to decreasing pore space, clastic reservoirs will receive less oil with depth and time, whereas carbonate reservoirs might be more receptive to oil with depth and time (i.e. secondary porosity).

(See Fig. 31 for evidence of early high temperature migration to clastic reservoirs at shallow depths and possibly late high to moderate temperature migration to carbonate reservoirs with time.)

* Carbonate pore space changes with time variable, unpredictable or unknown.

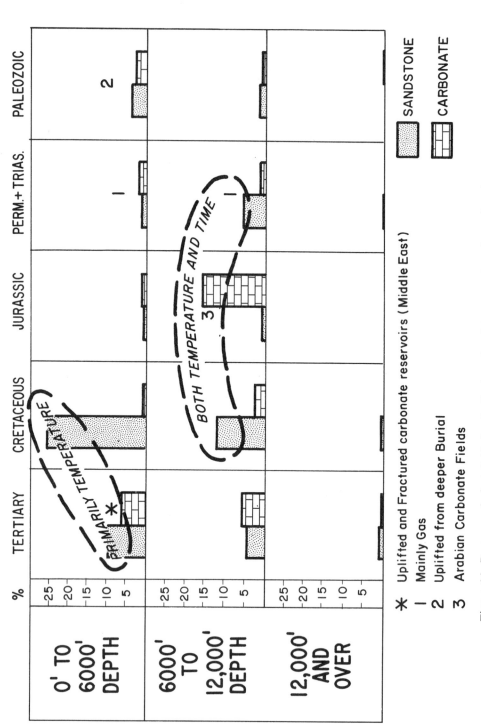

Figure 31. Reserves of giant fields by depth, type (sandstone versus carbonate) and age.

and depth of burial or "unroofing"). Time appears to be a modifying factor which aids the oil forming process (see Table 1 and Fig. 31) in a manner similar to that of temperature, but to a lesser degree.

The high amount of gas in the cratonic basins, much of it in shallow Paleozoic reservoirs, is often attributed to:

1. The possibility of long distance migration through a long period of geologic time to large regional arches located on the edges or in the center of basins (Panhandle-Hugoton, Hassi R'mel, Gronigen and West Siberian Arches).

2. The presence of widespread Carboniferous coal deposits providing a gas source (North Sea, Appalachia).

GENERAL CONCLUSIONS

1. Hydrocarbon formation and migration in clastic sequences is temperature related. High geothermal gradients appear to enhance petroleum generation and mobility to a greater degree than they adversely affect reservoir quality.

2. The geothermal gradients in sedimentary basins are generally related to the heat flow of a broader region. Some basins, however, may lose part of their heat by convective processes not recorded in the measurements of the gradient, and others show reduced geothermal gradients due to rapid sedimentation or to the insulating effect of overpressured shales at depth. Some basins or parts of basins are hot spots, anomalous in areas of lower heat flow.

3. The main types of basins show characteristic geothermal gradients, and may have been subjected to characteristic histories of heat flow. Individual differences between basins of the same type reflect differences in tectonic and sedimentational development.

4. Differences in the thermal gradients of basins, and in the history of such gradients during basin development, must play a major role in the history of petroleum generation, migration, and accumulation, and may determine the quantity of recoverable petroleum as well as the size and the depth of major accumulations. The low petroleum yield of many cratonic basins may be principally related to their low heat flow. The high yield of many basins of intermediate crustal type, such as basins of Types 6 and 7, associated with continental margins, may reflect the high heat flow of these areas (Fig. 30).

5. The temperatures in basins can, of course, become too high for

the retention of hydrocarbon accumulations, well before the rocks as such begin to show metamorphic effects. In the end, many petroleum-bearing basins fall victim to the high temperatures associated with orogeny. The basins least likely to do so are those of the stable cratonic interior, and the ones most threatened are those of the intermediate crust type—the very ones whose high heat flow has made them the most productive ones.

Viewed in this light, the cratonic basins appear like tank ships of mainly Paleozoic source and reservoir rocks, moored to Precambrian shields, and transported along with these shields to their present positions by sea floor spreading. The present intermediate crustal basins were developed either in front of or in the "wake" of the continent as a result of the sea floor spreading in post-Paleozoic time. As such their hydrocarbons were formed on their way to their present positions as they passed over the subcrustal bunsen burner of high heat flow which has been present along some of the continental margins. The flow of heat and related subsurface temperatures have enhanced the formation of giant fields in post-Permian basins. Conversely, many or most of the pre-Permian intermediate crustal basins have been destroyed by the thermally active Hercynian orogeny (i.e. concentric, accretionary belts around shield areas, Fig. 9). The Paleozoic basins of more mobile, continental-margin type were presumably richer in hydrocarbon accumulations, but largely fell victim to the Hercynian orogenies. Most of the prolific basins of high heat flow on present continental margins, formed in the Mesozoic-Cenozoic episodes of sea-floor spreading, will in turn be destroyed in the next round of orogenies.

6. Geothermal data are not available for many of the world's sedimentary basins, and, for most, the data are inadequate to determine the ultimate validity of some of the suggestions made in this study. General release of temperature data by the petroleum industry is to be desired, as is the systematic compilation of geothermal patterns round the world.

It is concluded that within the clastic facies of sediments high heat flow (within the limits of depth and time) result in greater than normal reserves of hydrocarbons, providing all the necessary geologic factors for oil accumulation are present. All basin types appear to show this general relation. Although the effects of heat flow can be demonstrated in each type of basin, Type 6 and 7 intermontane basins are the most dramatic in their demonstration of this relation.

REFERENCES

Athy, L. F., 1930. Density, porosity and compaction of sedimentary rocks, *Am. Assoc. Petroleum Geologists Bull.,* v. 14, no. 1, pp. 1–24.

Atwater, T. A., 1970. Implications of plate tectonics for the Cenozoic tectonic evolution of Western North America, *Geol. Soc. America Bull.,* v. 81, pp. 3513–3536.

Boldizar, Tibor, 1968. Geothermal data from the Vienna Basin, *Journal of Geophysical Research,* v. 73, no. 2, pp. 613–618.

Brooks, J. D., 1970. The use of coals as indicators of the occurrence of oil and gas, *Australian Petroleum Association Journal,* v. 10, no. 2, pp. 35–50.

————, 1970b. The reason why, *Broken Hill Proprietary Journal,* pp. 2–5.

Brooks, J. D. et al., 1971. The natural conversion of oil and gas in sediments in the Cooper Basin, *Australian Petroleum Exploration Association Journal,* pp. 121–125.

Burst, John F., 1969. Diagenesis of Gulf Coast clayey sediments and its possible relation to petroleum migration, *Am. Assoc. Petroleum Geologists Bull.,* v. 53, no. 1, pp. 73–93.

Chilingar, George V., and Larry Knight, 1960. Relationship between pressure and moisture content of kaolinite, illite and montmorillonite clays, *Am. Assoc. Petroleum Geologists Bull.,* v. 44, no. 1, pp. 101–106.

Delattre, I. et al., 1970. A provisional geothermal map of the Rhine Graben (Alsation part): from *Graben Problems*—Int. Upper Mantle Proj., Science Report no. 27, edited by Illies, Stuttgart.

Dott, R. H., Sr., and M. J. Reynolds, 1969. *Source book for Petroleum Geology,* Am. Assoc. Petroleum Geologists Memoir 5.

Emmons, W. H., 1931. *Geology of Petroleum,* New York, McGraw-Hill Book Company.

Epp, David et al., 1970. Heat flow in the Caribbean and Gulf of Mexico, *Journal of Geophysical Research,* v. 75, no. 29, pp. 5655–5669.

Erickson, Albert J., June 1970. The measurement and interpretation of heat flow in the Mediterranean and Black Sea, Cambridge, Massachusetts (Ph.D. Thesis—MIT).

Fotiadi, E. E. et al., 1970. Geothermal investigations in some regions of Western Siberia: in *Geothermal Problems,* Symposium, Madrid, Spain, 1969, Proc. Tectonophysics, v. 10, no. 1–3, pp. 95–101.

French, R. W., April 11, 1939. Geothermal gradients in California oil wells: API Directory of Production, Los Angeles, Calif.

Gardett, Peter H., 1971. Petroleum potential of the Los Angeles Basin, California. In *Petroleum Provinces of the United States—Their Geology and Potential,* Am. Assoc. Petroleum Geologists Memoir no. 15, vol. 1, pp. 298–308.

Griffin, G. M. et al., 1969. Geothermal gradients in Florida and southern Georgia, *Am. Assoc. Petroleum Geologists Bull.,* v. 53, no. 9, pp. 2037.

Gupta, M. L. et al., 1970. Terrestial heat flow and tectonics of Cambay Basin, Gujarat State (India), *Tectonophysics*, v. 10, pp. 147–163.

Halbouty, M. T. et al., 1970. Factors affecting formation of giant oil and gas fields and basin classification, Part II, *Am. Assoc. Petroleum Geologists Memoir* 14.

Hamilton, W., 1969. Mesozoic California and the underflow of Pacific mantle, *Geol. Soc. America Bull.*, v. 80, pp. 2409–2430.

Hamilton, W., 1970. The Uralides and the motion of the Russian and Siberian platforms, *Geol. Soc. America Bull.*, v. 81, pp. 2553–2576.

Hanel, R., 1970. Interpretation of the terrestrial heat flow in the Rhine Graben, In *Graben Problems,* Int. Upper Mantle Proj., Stuttgart.

Hedberg, Hollis, D., 1926. The effect of gravitational compaction of the structure of sedimentary rocks, *Am. Assoc. Petroleum Geologists Bull.*, v. 10, pp. 1035–1072.

——, 1936. Gravitational Compaction of clays and shales, *American Journal Sci.*, v. 31, pp. 241–287.

——, 1954. World Oil Prospects: from a geologic viewpoint, *Am. Assoc. Petroleum Geologists Bull.*, v. 38, no. 8, pp. 1714–1724.

——, 1970. Continental Margins: from viewpoint of the petroleum geologist, *Am. Assoc. Petroleum Geologists Bull.*, v. 54, no. 1, pp. 3–34.

Hunt, J. M., Geochemical data on organic matter in sediments, *Int. Sci. Oil Conf. Proc.*, Budapest, preprint.

Jam, L., Parke A. Dickey, and Eysteinn Tryggvason, 1969. Subsurface temperature in south Louisiana, *Am. Assoc. Petroleum Geologists Bull.*, v. 53, no. 10, pp. 2141–2149.

Jones, Paul H., July 1969. Hydrodynamics of geopressure in the northern Gulf of Mexico basin, *Journal of Pet. Tech.*, pp. 803–810.

Karig, Daniel E., 1971. Origin and development of marginal basins in the Western Pacific, *Journal of Geophysical Research,* v. 76, no. 11, pp. 2542–2560.

Karstev, A. A. et al., 1971. The principal stage in the formation of oil: Preprint 8th World Petroleum Congress, Moscow, Panel Discussion 1, p. 1–17, Elsevier Pub. Co. Ltd.

Kay, M., 1951. North American geosynclines, *Geol. Soc. America Mem.* 48, p. 143.

Kilkenny, John E., 1971. Future petroleum potential of region 2, Pacific Coastal states and adjacent continental shelf and slope. In *Future Petroleum Provinces of the United States,* Am. Assoc. Petroleum Geologists Memoir 15, v. 1, pp. 170–177.

King, P. B., comp., 1969. *Tectonic map of North America,* U.S. Geol. Survey, 1:5,000,000.

Klemme, H. D., March 1, 8 and 15, 1971. What giants and their basins have in common, *Oil and Gas Journal.*

——, October 1971 (b). Trends in basin development: Possible economic implications, *World Petroleum.*

Landes, Keneth H., 1967. Eometamorphism, and oil and gas in time and space, *Am. Assoc. Petroleum Geologists Bull.,* v. 51, no. 6, pp. 828–841.

Lee, W. H. K., ed., 1965. *Terrestial Heat Flow:* American Geophysical Union, Monograph 8.

Levorsen, A. I., 1967. *Geology of Petroleum,* San Francisco, Freeman and Co., Chap. 9, p. 415.

Lewis, C. R. and S. C. Rose, September 1969. A theory relating high temperatures and overpressures, Soc. of Petroleum Eng. AIME, 44th Annual Fall Meeting, Denver, Colorado.

Lubimova, E. A. 1969. Thermal History of the Earth, *The Earth's Crust and Upper Mantle,* American Geophysical Union Monograph 13, Chap. 10.

Makarenko, F. A. et al., 1968. Geothermal field on the U.S.S.R. territory: XXIIIM Int. Geol. Congress, v. 5, p. 67–73.

Mayuga, M. N., 1970. Geology and development of California's giant-Wilmington oil field, *Geology of Giant Petroleum Fields,* Am. Assoc. Petroleum Geologists Memoir 14, pp. 158–184.

Maxwell, John C., 1964. Influence of depth, temperature and geologic age on porosity of quartzose sandstone, *Am. Assoc. Petroleum Geologists Bull.,* v. 48, no. 5, pp. 697–709.

Meidava, Tsvi, and R. W. Rex, 1970. Geothermal exploration in Imperial Valley, *Am. Assoc. Petroleum Geologists Bull.,* v. 54, no. 3, p. 560.

Moiseyenko, V. I. et al., 1970. Ein Versuch der Bestimmung des Erdwärmeflusses in Flachbohrungen. In *Geothermal Problems,* Symposium, Madrid, Spain 1969, Proc. Tectonophysics, v. 10, no. 1–3 (Special Issue), pp. 89–94.

Nagle, H. E., and E. S. Parker, 1971. Future oil and gas potential of onshore Ventura Basin, California. In *Future Petroleum Provinces of the United States—Their Geology and Potential,* Am. Assoc. Petroleum Geologists Memoir 15, v. 1, pp. 254–297.

Oxburgh, E. R., and D. L. Turcotte, 1970. Thermal structure of island arcs, *Geol. Soc. America Bull.,* v. 81, pp. 1665–1688.

Peive, A. V. ed., 1970. Heat regime of the U.S.S.R. interiors, translation, v. 218, NAUKA Publ., Moscow.

Perry, E. A. Jr. and John Hower, October 1972, Late stage dehydration in deeply buried pelitic sediments: Am. Assoc. Petroleum Geologists Bull., v. 56, no. 10, p. 2013–2021.

Philippi, G. T., 1965. On depth, time and mechanism of petroleum generation, *Geochimica et Cosmochimica Acta,* v. 29, pp. 1021–1049.

Pusey, W. C., 1973. How to evaluate potential gas and oil source rocks, *World Oil,* v. 176, no. 5, pp. 71–75.

Sass, John H., 1971. The earth's heat and internal temperature. In *Understanding the Earth,* (I. G. Gass, Peter J. Smith, R. C. L. Wilson, eds.), Sussex, England, Artemis Press, Chap. 5, pp. 80–87.

Sass, J. H. et al., 1971. Heat flow in western United States, *Journal of Geophysical Research,* v. 76, no. 26.

Scholz, C. H. et al., 1971. Late Cenozoic evolution of the Great Basin, western United States, as an ensialic interarc basin, *Geol. Soc. America Bull.,* v. 82, pp. 2979–2990.

Schoeppel, Roger J. et al., October 1, 1970. Geothermal survey of North America, second annual report 1970: unpublished Am. Assoc. Petroleum Geologists committee report.

Sclater, J. G., and J. Francheteau, 1970. The implications of terrestial heat flow observations on current tectonic and geochemical models of the crust and upper mantle of the earth, *Geophysical Journal,* v. 20, no. 5, pp. 509–542.

Tamrazyan, G. P., 1971. Siberian continental drift, *Tectonophysics,* v. 11, pp. 433–460.

Timko, D. J., and W. H. Ferth, August 1971. Relationship between hydrocarbon accumulation and geopressure and its economic significance, *Journal of Pet. Tech.,* pp. 923–933.

Tissot, B. et al., 1971. Origin and evolution of hydrocarbons in early Toarcian shales, Paris basin, France, *Am. Assoc. Petroleum Geologists Bull.,* v. 55, no. 12, pp. 2177–2193.

Uspenskaya, N. Yu., 1967. Principles of oil and gas territories subdivisions and the classification of oil and gas accumulations, 7th World Petroleum Congress, Mexico, *Proceedings,* Amsterdam, Elsevier Pub. Co., v. 2, pp. 961–969.

von Herzen, R. P. et al., 1970. Heat flow between the Caribbean Sea and the mid Atlantic ridge, *Journal of Geophysical Research,* v. 75, no. 11, pp. 1973–1984.

Vredenburgh, L. D., and E. S. Cheney, 1971. Sulfur and carbon isotopic investigations of petroleum, Wind River basin, Wyoming, *Am. Assoc. Petroleum Geologists Bull.,* v. 55, no. 11, pp. 1954–1975.

Walker, Kenneth R., 1964. Influence of depth, temperature and geologic age on porosity of quartzose sandstone (discussion), *Am. Assoc. Petroleum Geologists Bull.,* v. 48, no. 12, pp. 1945–1946.

Watanobe, Teruhiko et al., 1970. Heat flow in the Philippine Sea, *Tectonophysics,* v. 10, pp. 205–224.

Weeks, Lewis G., 1952. Factors of sedimentary basin development that control oil occurrence, *Am. Assoc. Petroleum Geologists,* v. 36, no. 11, pp. 2071–2124.

———, 1958. *Habitat of Oil,* Am. Assoc. of Petroleum Geologists.

———, 1965. World offshore petroleum resources, *Am. Assoc. Petroleum Geologists Bull.,* v. 49, no. 10, pp. 1680–1693.

———, 1971. Marine geology and petroleum resources. Preprint, 8th World Petroleum Congress, Moscow, Panel Disc. 2, pp. 1–15.

Welte, Dietrich H., 1965. Relation between petroleum and source rock, *Am. Assoc. Petroleum Geologists Bull.,* v. 49, no. 12, pp. 2246–2268.

Zierfuss, H., 1969. Heat conductivity of some carbonate rocks and clayey sandstones, *Am. Assoc. Petroleum Geologists Bull.,* v. 53, no. 2, pp. 251–260.

Having opened with a perspective by an academic scientist, it is fitting that we should close with one by a man of Affairs: John Moody examines the geological setting of the 198 known giant oil fields, each expected to yield more than 500 millions barrels, which in aggregate contain some 67% of the world's known reserves. Some of the patterns revealed in this study find ready scientific explanation, while others do not, and all afford food for thought.

Distribution and Geological Characteristics of Giant Oil Fields

J. D. MOODY[1]

When the first giant oil field, the Bradford field of Pennsylvania, was discovered in 1865, the world's population was only about one-third as large as it is now, and market demand for oil was rather slow in developing. Some 40 years ago, the discovery of the east Texas field, now estimated to have an ultimate recovery of 5.6 billion barrels, flooded the market and broke the price of oil in the United States. This was a dramatic example of the effect of finding a giant oil field then. Today, the recent discovery of the Prudhoe Bay field, two to three times as large, is merely a welcome addition to the nation's supply.

We define a giant oil field as one whose ultimate recovery under known technology will equal or exceed 500 million barrels. We have considered as a single field an aggregate of contiguous, overlapping, or superimposed reservoirs whose combined ultimate will satisfy this criterion. Please note that gas, either associated or nonassociated with crude oil, does not enter into this discussion.

[1] Mobil Oil Corporation, New York, New York.

In the years that have followed the discovery of Bradford, the exploration for oil has become quite sophisticated, quite scientific, and very competitive. The world has consumed some 240 billion barrels of oil, about 16 billion barrels in 1970 alone. This ever-increasing demand and the enormous quantities of oil being consumed make it increasingly necessary to concentrate our efforts on the search for giant oil fields. From now on, we can hope to supply the anticipated demand only by findings more giants.

In studying the geology and geography of the giant oil fields of the world in some detail, we have uncovered some unexpected (to us, anyway) features. For example, it is interesting to observe that, by ten-year increments, more giant fields were found in the last decade than in any previous ten-year period, as illustrated by Figure 1. It appears, however,

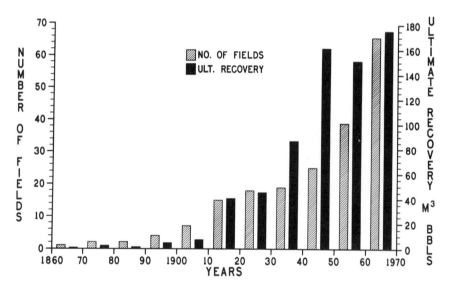

Figure 1. Time of discovery of giant oil fields, total world.

that the average size of the discoveries is decreasing somewhat. The 198 fields that constitute this study contain an estimated ultimate recovery of about 675 billion barrels. This amounts to about 67% of the estimated world total, and includes cumulative production as well as remaining reserves.

Figure 2 is an analysis of the geologic ages of the producing reservoirs of the giant fields and emphasizes clearly the prolific nature of the Tertiary and Mesozoic reservoirs. The Mesozoic, mostly because of the very large accumulations in the Middle East, holds the highest rank. The relationship is so striking that Mesozoic and Tertiary sedimentary areas of the world must be considered as the prime exploration targets. One possible explanation of this important relation has to do with the worldwide Hercynian tectonism. Widespread evaporites and red beds of

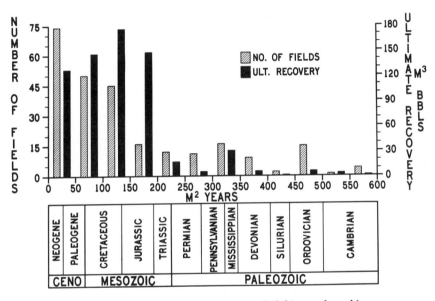

Figure 2. Age of reservoirs of giant oil fields, total world.

Permian and Triassic ages provide evidence that a worldwide postorogenic thermal maximum must have followed on the heels of the Hercynian orogenies. Such a postorogenic thermal maximum might have enhanced generation of oil from then-existing source rocks; it might have mobilized pre-existing oil accumulations and caused them to remigrate; it might have destroyed other previously existing oil accumulations; and it might have enhanced the bio-environment in immediately subsequent source beds of Jurassic and lower Cretaceous age, resulting in postorogenic source rock enrichment.

Figure 3 shows average reservoir depth of the main pay of the giant

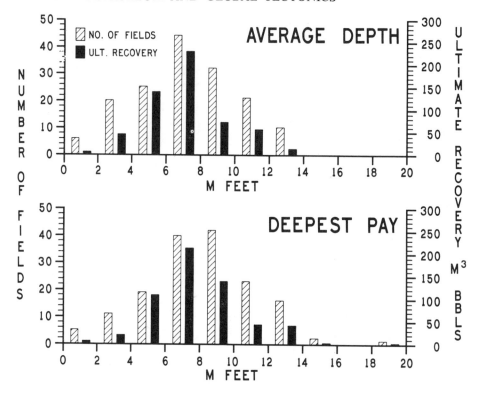

Figure 3. Depth of reservoirs of giant oil fields, total world.

fields included in our study. It is clear that major accumulations of oil are not common below about 12,000 ft.

When we examine the nature of the reservoirs of the giant fields, we find that there are 50% more sandstone than carbonate reservoirs, as shown in Figure 4. However, the carbonates contain almost as much total oil, which indicates that the carbonate giants on the average are much larger than the sand giants. Reservoir rock other than sandstone or carbonate is insignificant.

In looking at the primary mode of trapping involved in the giant fields, Figure 5 shows that anticlines greatly outnumber all other trapping mechanisms. It may reasonably be argued that this is so because we can find anticlines more readily than other types of traps. However, it also may be that anticlines are the most efficient type of trapping mechanisms.

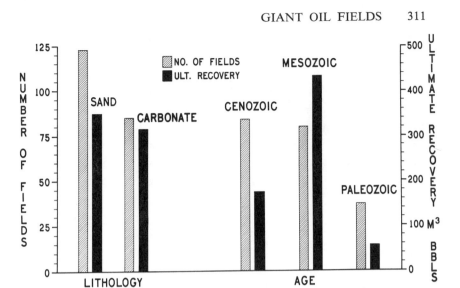

Figure 4. Lithology of reservoirs of giant oil fields, total world.

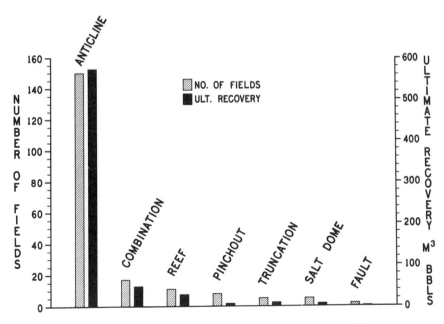

Figure 5. Mode of trapping of giant oil fields, total world.

Much attention has been given to the problems of deciphering the geologic history of giant fields. Of interest are the dating of structural growth, and the dating of oil formation, migration, and accumulation. For many of the fields, especially some of the older ones, adequate information is often lacking. We decided to see what, if any, relationship exists between giant accumulations and significant breaks in sedimentation. Figure 6 shows that the noninvolvement of significant unconformities is a favorable condition for the formation of giant oil fields; however, this tabulation (as are all the others) is heavily weighted by the

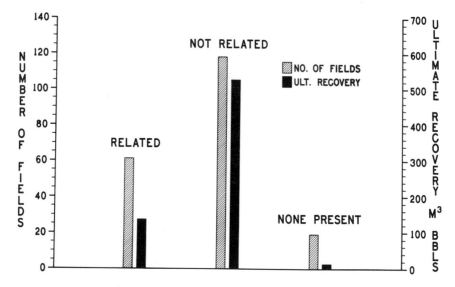

Figure 6. Unconformities in relation to giant oil fields, total world.

giant fields occurring in the Middle East. Significant unconformities are related to the giant fields outside the Middle East more often than not, and are involved in the accumulation and trapping mechanism of a majority of these fields.

In a similar vein, Figure 7 shows that, although not essential, the presence of evaporites as a sealing agent is a favorable circumstance. Fields with evaporite sealing beds hold, on the average, much more oil than those without such caps. Here again, the giant fields of the Middle East illustrate this principle.

Figure 8 clearly shows that the shelf areas are far and away the pre-

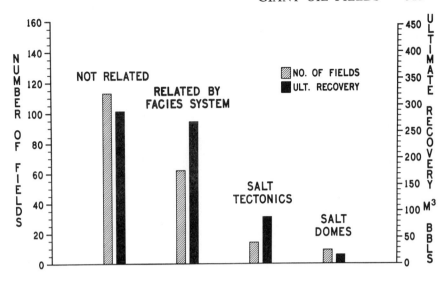

Figure 7. Evaporites in relation to giant oil fields, total world.

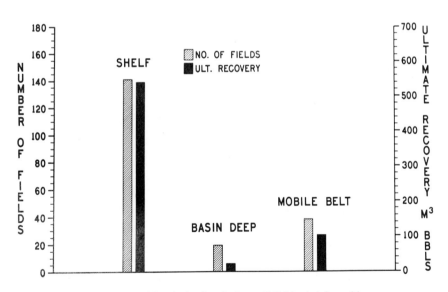

Figure 8. Position in basin of giant oil fields, total world.

ferred location for the development of giant oil fields. This is explainable by the favorable conjunction of reservoir beds, source materials, and rapid deposition in this type of environment. Any preservation of accumulations is probably greater on the shelf side of a basin. There are undoubtedly other factors not yet recognized.

During the first 100 years of exploration, some 1,000 billion barrels of ultimately recoverable oil have been found. About 67% of this amount is contained in the 198 giant fields of this study. There are probably a dozen or more giants that have been found within the past few years for which data are presently not available, and hence are not included in this study. Other recently discovered giants have not yet been recognized as such. Figure 9 shows those basin areas in which the known giant fields are found, and the number in each basin. The Middle East, of course, has the largest number and the greatest amount of oil.

As the need for oil has increased, so the quality of our exploration has improved. We are finding more giant fields now than in the past although their average size may be somewhat smaller. This, with the exception of the western Siberian basin of the USSR and the Persian Gulf basin, would not be true were it not for the recent great expansion of offshore exploration. Indeed, the offshore areas of the world offer the best remaining hunting grounds for giant fields. As we have shown above, those sedimentary areas with adequate thicknesses of Tertiary and Mesozoic rocks that were deposited in a shelf environment are prime targets. The continental shelves are the largest remaining relatively unexplored domains of these sediments, and the current intense worldwide exploration activity on the shelves testifies to the oil industry's recognition of their prospectiveness.

I have some difficulty in trying to relate giant oil fields to the "New Global Tectonics" per se. Indeed most of the 198 giant oil fields that are the subject of this study were found before the advent of "New Global Tectonics." Apparently a judicious combination of petroleum geology, exploration history, and normally evolving technology was sufficient to uncover the first trillion barrels of earth's oil resources. But the discovery of the second trillion barrels will be considerably more difficult, and searchers for oil will need all the help they can get.

However, if for "New Global Tectonics" we substitute "Post-Hercynian Global Tectonics," thereby assigning the term "new" to a geological time-scale context, we find a rather striking correlation as shown in Figure 10. In this context, the bulk of the world's giant fields, and

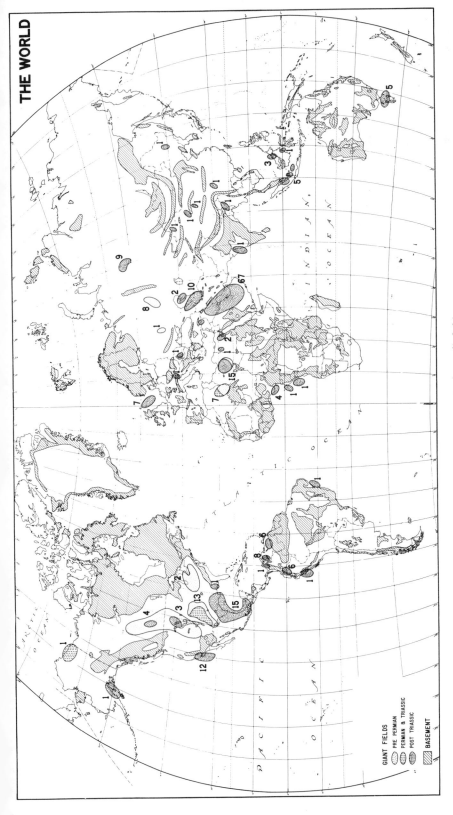

Figure 9. Location of giant oil fields.

the bulk of the world's oil, can be seen to be reservoired in rocks of post-Triassic age. And it is generally accepted that the various tectonic events, which taken collectively constitute the "New Global Tectonics," span the last 180 million years of geologic time—i.e., post-Triassic time. There can be little doubt that the post-Triassic tectonism of the new global tectonics is inherited from the pervasive orogenic episodes of the late Paleozoic Hercynian deformation. As suggested earlier in this paper,

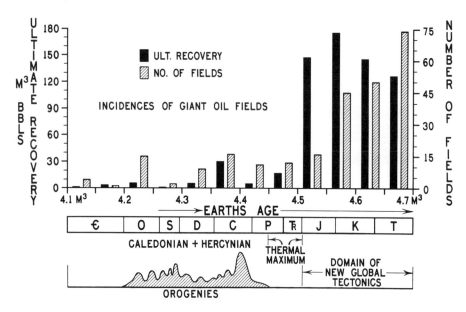

Figure 10. Post-Hercynian concentration of giant oil fields, total world.

Hercynian orogeny was accompanied and/or followed by major changes in the earth's heat flow and energy balance—post-Hercynian geothermo-dynamics must surely be important to the new global tectonics, and may even be a principal causative link between the new global tectonics and the occurrence of giant oil fields.

Figure 11 shows the relation between areas of basins containing giant fields, and crustal plates of the new global tectonics, and Figure 12 shows the relation between those areas containing giant oil fields and major elements of the earth's regmatic shear pattern. These two geotectonic syntheses are based on quite different concepts, but both contem-

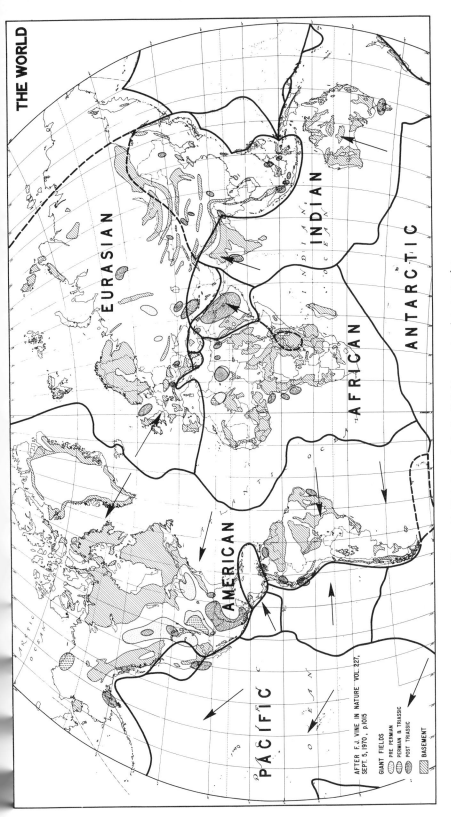

Figure 11. Relation of giant oil field areas to plate tectonics.

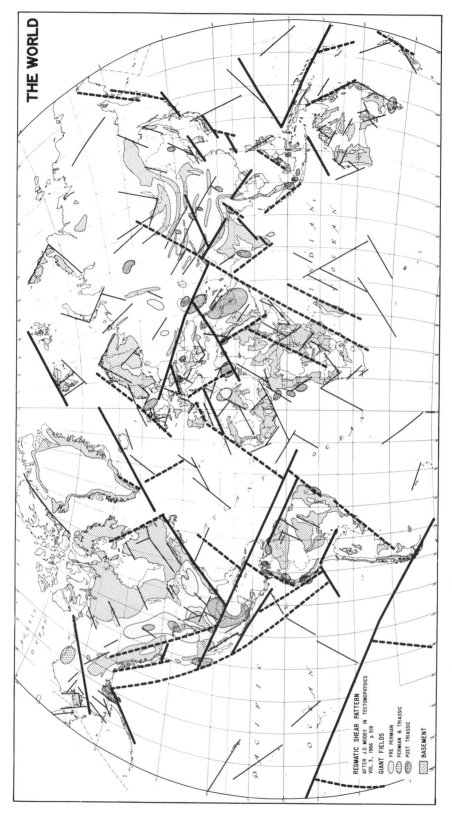

Figure 12. Relation of giant oil field areas to regmatic shear pattern.

plate crustal fragmentation into a mosaic of plates or blocks. In the words of Professor L. DeSitter (1956), ". . . geotectonical synthesis of all structural knowledge, is at the same time in a certain sense its [structural geology's] most unsatisfactory branch, because so many totally unknown and untested properties of the deeper crust, the mantle, and the core of the earth are involved. It is the territory of the most unchecked and flamboyant fantasy, built up by piling hypotheses on theories, shored by very rare and thin reeds of geophysical facts."

Since DeSitter wrote these words, many more reeds have been added and incorporated into both of the syntheses represented by Figures 11 and 12. And the reeds, if not the syntheses, must be accommodated in our geotectonic understanding. In 1968 plate tectonic theory recognized six "plates," in 1972 the number had grown to 20. It seems possible that the two different schemes of global tectonics represented by Figures 11 and 12 are converging, at least so far as crustal disruption into a mosaic of plates or blocks is concerned. Hopefully our understanding of the earth, its history, and its processes will continue to increase.

Although we are convinced that no one can foretell the future accurately, it is often instructive and useful to indulge in speculative estimates. Among the several estimates of potential oil yet to be found in the world is our own of 800–900 billion barrels. Let us assume that 75% of this estimated amount will be found in giant fields. If this should come to pass, we would expect that there should be on the order of 200–300 giant fields remaining to be found, depending upon what the average size of the undiscovered giants be assumed to be.

Searching for these undiscovered giant oil fields will be a challenging task and finding them will be a rewarding experience. May we all have our deserved share of success.

REFERENCES

DeSitter, L. J., 1956. *Structural Geology*, McGraw-Hill Series in the Geological Sciences, 552 pp.

Dewey, John F., 1972. Plate tectonics, *Scientific American,* v. 226, no. 5, pp. 56–68.

Gurari, F. B., 1971. Oil and gas of Western Siberia, prospects and problems, *Priroda,* no. 1, January 1971, pp. 16–23 (Eng. trans., *Review of Sino-Soviet Oil,* Arlington, Va., v. 6, No. 11, April, 1971, pp 13–20).

Halbouty, M. T., A. A. Meyerhoff, R. E. King, R. H. Dott, Sr., H. D.

Klemme, and T. Shabad, 1970. World's Giant Oil and Gas Fields, Geologic Factors Affecting Their Formation and Basin Classification. Memoir No. 14 Am. Assoc. Petroleum Geologist, 1970.

King, R. E., 1969. Oil and gas exploration and production in the Soviet Union. In *Exploration for Petroleum in Europe and North Africa,* London Institute Petroleum, pp. 181–190, 1969.

Klyuchnikov, N. I., and A. G. Bilbulatov, 1964. Principal results of oil and gas prospecting and exploration in Bashkiria during the first five years of the seven year plan and prospects for further development, *Geologiya Nefti i Gaza,* v. 8, no. 10, 1964 (Eng. trans. in *Petroleum Geology,* McLean, Va., July 1970, pp. 543–548.

Knebel, G. M., and G. Rodriguez-Eraso, 1956. Habitat of some oil, *Am. Assoc. Petroleum Geologists Bulletin,* v. 40, no. 4, pp. 546–561.

Moody, J. D., 1966. Crustal shear patterns and orogenesis, *Tectonophysics,* v. 3, no. 6, pp. 479–522.

Moody, J. D., J. W. Mooney, and J. Spivak, 1970. Giant Oil Fields of North America, *Geology of Giant Petroleum Fields,* Memoir No. 14, Am. Assoc. Petroleum Geologists.

Shashin, V. D., 1971. A Round Million. Interview with Minister of the Oil Industry of the USSR. *Literaturnaya Gazeta,* January, 1971, P. 10. (Eng. trans. in *Review of Sino-Soviet Oil,* Arlington, Va., v. 6, no. 9, February 1971, pp. 5–7.)

Vine, F. J., 1970. The geophysical year, *Nature,* v. 227, pp. 1013–1017.

Weeks, L. G., 1952. Factors of sedimentary basin development that control oil occurrence, *Am. Assoc. Petroleum Geologists,* v. 36, no. 11, pp. 2071–2124.

Yenikeyev, P. N., P. T. Kozlov, and P. Ye. Yavkin, 1965. Oil-gas resources of Middle Asia and their prospects, *Geologiya Nefti i Gaza,* v. 9, no. 2. (Eng. trans. in *Petroleum Geology,* McLean, Va., February 1970, pp. 67–69.)

Epilogue

Academic science and its application to human affairs move along in an interlinked manner. Today's search for fuel resources has at its command a much greater armory of data-gathering and data processing devices than that of yesteryear; but the petroleum geologist is also concerned with a much broader range of questions. His inquiry no longer stops with the availability of reservoir rocks and the presence of stratigraphic or tectonic traps, but extends to questions of the chemical environments of deposition and diagenesis, the origin and history of basins, their temperature and pressure regimes influential in hydrocarbon generation, and the movements of fluids throughout basin history.

Plate tectonic theory, sweeping the field of geology as a whole, appears to many of us as a revolutionary advance in our understanding of the earth (while some consider it as a major aberration). To the editors its greatest appeal lies in its universality: it appears to explain and relate a large number of previously unexplained and unrelated observations. The papers of this symposium suggest that it holds the keys to many problems vital to the understanding of the generation and distribution of fossil fuels: the origin of basins, sources of sediment, open versus restricted circulation, prediction of basin geometry, basin relationships on opposed continental margins, and thermal regimes within basins. Thus even at its present, early development plate tectonics appears to be offering significant aid to the search for fuels. Modern society is primarily dependent on fossil fuels; its future may hinge on whether these fuels can continue to be found and produced until technology reaches levels at which other less exhaustible forms of energy become available. We hope that the conference has already contributed to the strategy and tactics of petroleum exploration, and that this book, reaching a wider audience, will contribute further toward this end.

At the same time, the discoveries of industry are and will be feeding large quantities of information back into academic concerns of how and why plates are moving, to complete the symbiosis of Industry and Academe, which has ever played such an important role in our science, and which is so eloquently expressed in the career of Hollis D. Hedberg.

THE EDITORS

LIBRARY OF CONGRESS CATALOGING IN PUBLICATION DATA
Main entry under title:

Petroleum and global tectonics.

Revised versions of papers presented at the 109th
meeting of the Princeton University Conference held on
March 10 and 11, 1972.
    1. Petroleum—Geology—Congresses. 2. Plate
tectonics—Congresses. I. Fischer, Alfred G., ed.
II. Judson, Sheldon, ed. III. Princeton University
Conference.
TN870.5.P46 1975   553'.28   72-9946
ISBN 0-691-08124-7
ISBN 0-691-08128-X pbk.